HOW
THEY DO IT

HOW
THEY DO IT

Robert A. Wallace

WILLIAM MORROW AND COMPANY, INC.
New York   1980

To my students—
Who first reminded me that I didn't know either

Library of Congress Cataloging in Publication Data

Wallace, Robert Ardell, 1938-
  How they do it.

"Morrow quill paperbacks."

  Bibliography: p.
  Includes index.
  1. Sexual behavior in animals. I. Title.
[QL761.W28 1980b]   591.1'6   80-17605
ISBN 0-688-03718-6
ISBN 0-688-08718-3 (pbk.)

Printed in the United States of America

First Morrow Quill Paperback Edition

1 2 3 4 5 6 7 8 9 10

BOOK DESIGN BY MICHAEL MAUCERI

Photo Acknowledgments

Parakeets, p. 2. © Toni Angermayer/National Audubon Society/ Photo Researchers

Pigs, p. 20. Institute for Sex Research

Cats, p. 28. © Hans Reinhard/Bruce Coleman Inc.

Earthworms, p. 36. © Laurence Pringle/NAS/Photo Researchers

Seahorses, p. 48. Keystone

Newts, p. 54. © Lynwood M. Chace/NAS/Photo Researchers

Elephants, p. 57. © Marc Boulton/NAS/Photo Researchers

Snails, p. 64. © L. & D. Klein/NAS/Photo Researchers

Praying Mantises, p. 68. © Jen and Des Bartlett/NAS/Photo Researchers

Alligators, p. 75. © Robert C. Hermes/NAS/Photo Researchers

Tarantulas, p. 88. © Jen and Des Bartlett/Bruce Coleman Inc.

Snakes, p. 99. © Tom McHugh/NAS/Photo Researchers

Damselflies, p. 106. © Russ Kinne/NAS/Photo Researchers

Kangaroos, p. 112. Institute for Sex Research

Deep Sea Fish, p. 116. © Peter David/NAS/Photo Researchers

Squirrel Tree Frogs, p. 119. © S. McKeever/NAS/Photo Researchers

Common Frogs, p. 121. © Jane Burton/Bruce Coleman Inc.

Rhinoceroses, p. 125. © Jacana/The Image Bank

Grunion, p. 129. © Tom McHugh/NAS/Photo Researchers

East African Mosquito Fish, p. 129. © Jane Burton/Bruce Coleman Ltd.

Hippopotamuses, p. 133. © Christian Polit/Institute for Sex Research

King Penguins, p. 137. © Francisco Erize/Bruce Coleman Inc.

South American Forest Tortoises, p. 140. © Tom McHugh/NAS/ Photo Researchers

Chameleons, p. 147. © Frederick Ayer/NAS/Photo Researchers

Bacteria, p. 150. © T. F. Anderson (Institute for Cancer Research), and E. L. Wollman and F. Jacob (Pasteur Institute, Paris)

Flamingos, p. 153. © C. Ray/NAS/Photo Researchers

Camels, p. 157. Institute for Sex Research

The Publisher has made every effort to identify the source of all photographic material used in this book and to obtain permission to reproduce each photograph. Nonetheless, if we have failed to do so in any case, the appropriate persons are invited to contact the Publisher.

CONTENTS

8 HOW THEY DO IT

ABOUT THIS BOOK

This book tells how they do it. "They," in this case, are the other animals. It doesn't take much imagination to suspect that "it" is the sexual act—copulation. Of course, a great deal has been written about how *we* do it, both in aseptic and clinical terms, and vulgarly, to arouse our prurient interest—whatever *that* is. But while we agonize over the techniques and moral questions of our sex, the other animals seem to suffer no such burdens, accepting their sexual roles without those pangs of guilt, so perhaps their sex is of the less tainted sort. In any case, they're out there doing it right now in ways which are often intriguing and bizarre, or even startling.

As you wend your way through these pages you may be surprised, and perhaps relieved, to learn that animals, too, have their sexual "problems." Some of us have always suspected that only brainy and introspective humans could come up with problems relating to so simple and pleasurable an experience as sex. But if you believe that, you just don't know enough about the other animals. Female dogs in full heat apparently reject certain suitors on the basis of "personality." Homosexuality is rampant among bedbugs. Geese form ménages à trois. Rhinoceroses could be accused of sadomasochism. And skunks, orangutans, and lobsters rape. Indeed, there is hardly an aberrant sexual pattern found among our own species that is not shared by some other animal. No matter how sex is accomplished, however, the imperative remains the same: leave your kinds of chromosomes or suffer

9

genetic death. Basically, then, we are interested in how animals go about leaving their kinds of chromosomes in the next generation.

We almost never see other animals actually engaged in copulation in spite of the fact that we know it's always going on around us. Animals seem to keep their sexual lives a rather private matter. Sometimes, however, we may see more of it than we wanted. We have probably all been chagrined in our youth to have a romantic stroll shattered beyond all repair by walking up on a pair of dogs frantically copulating away. We have also seen them "tied" together and may have wondered about it. How does it happen? And why? And why does copulation in certain species sometimes result in agonizing death? How do those little females with no vaginas do it? And what about porcupines?

So this book will seek to answer some of those questions. It is a compilation of decades of work by a close fraternity of specialists, the behavioral zoologists. These dedicated souls have spent untold hours surreptitiously watching animals do it and taking very careful notes. Their observations have been translated into technical jargon and published in arcane journals, where only fellow specialists can find and decipher them. The conventions of this craft are very rigid. Descriptions must be very dry and couched in the precise language of the specialists; above all, the behavioral zoologist must never speculate about the animal's emotions or motives and must never draw analogies between animal and human behavior—that's called *anthropomorphizing* and is grounds for expulsion from the club.

I've written this book for fun and not for any scientific purpose. Let it be clearly understood: I claim no redeeming scientific or social merit for this exercise except to the extent that entertainment is itself redemptive. I believe my facts, however, are accurate and could legitimately lead one to further investigations. Actually, why sex exists at all, and in such varied hues, is a question yet to be satisfactorily answered.

So here I'll describe the sex lives of only certain species as representative of their groups. My accounts are drawn from my own twenty years' experience as a professional zoologist and from the kinds of scientific sources mentioned above. In certain cases, I've built a composite description from a number of sources. I've tried to avoid complex and overly scientific language. And I have allowed myself a certain amount of anthropomorphizing, again, just for fun. I may say an animal is angry, horny, or interested; but, of course, we don't really know what (if anything) an animal thinks or feels. And human feelings may have no counterparts among other animals. As far as what actually *happens*, though, as I said, I've tried to be accurate. If you want to know more about any aspect of what I've written here (or if you simply don't believe me), I've included a list of my sources at the end of the book.

I suspect that, to some degree, our lack of information on a fascinating subject stems from a lingering Victorian code. Sex has just not been a fit topic—at least not the way we will consider it. Hopefully, though, the times have changed so that at last we can explore the strange, bizarre—and sometimes downright funny—ways the other animals which share our planet . . . do it.

OCTOPUSES

An Arm and a Leg

Mating among octopuses can be a savage affair. In fact, some observers have also reported that the female *always* leaves the male mortally wounded and that he leaves the scene only to die in some shadowy cavern. We now know these accounts are not entirely accurate, but what really happens is strange enough.

Although the vigorous octopus is related to the uninspiring clam, it is undoubtedly even smarter than some vertebrates. It is also very excitable and has sharp eyes. Thus an intelligent, soft-bodied male eagerly signals to his prospective mate by visual displays, and rather startling ones at that. When he is in "the mood" (which usually means as soon as he spots a female), his normally grayish body suddenly becomes *striped*, and he literally becomes horny as two fleshy knobs rise from the top of his head! The stripes somehow seem to shimmer as he boldly swims right up to the larger, and hence more dangerous, female. As he draws near, within reach of her powerful tentacles, she normally holds her peace. Waiting. So far so good.

Perhaps, though, he is being hasty. At this point, he may stop and curl back his arms, presenting his undersurface and especially several oversized suckers on two of his arms. Another display. It is imperative that the male establish his sexual credentials with the female so that he triggers sexual, and not predatory, behavior.

She still hasn't responded, and so he grows bolder. It's

time to copulate, and now the story grows stranger. One reason is because he has no penis. He gets around this small embarrassment, however, by using his arm. It turns out that one of his eight appendages is a bit different from the others, and this one can act as a kind of copulatory organ.

Earlier, the sperm were packaged deep in his body by an extremely complicated process and are now enclosed in a number of tough little packets. Once the packets are inside the female, they will rupture and release their sperm. The problem, of course, is how to get them there.

At this point, the male swims above the female and begins caressing her with his other arms, touching her lightly here and there and softly stroking her back. If she allows this, she will probably allow him to copulate with her. The excitement of both animals can be seen in the lively color changes which sweep over their bodies. Suddenly he reaches toward her and quickly slips his reproductive arm into her siphon—her breathing tube.

This is the moment of truth in a number of respects, because she may respond in either of two ways, one of which spells trouble. If he continues to behave himself, she is likely to lie there peacefully but breathing hard, while the tip of his arm probes deep into the cavity under her great head. The French once reported that their octopuses *always* do it gently, the male behaving with *une certaine delicatesse*. Recently, however, other observers have found somewhat more gauche behavior on his part. The male, even a French one, may not always approach the female so gently. In some cases, he makes a jet-propelled leap toward the female and pins her down, covering her with the webs between his writhing arms. This is not so smart because she fights back, and now we have what is called "violent" copulation. He quickly learns that she can be a very tough opponent. As they grapple, powerful tentacles intertwine and suckers pull strongly at soft flesh. One animal may even wrap an arm around the mantle cavity of the other, cutting off respiration entirely and killing its mate. They may even rip at each other with their sharp parrot beaks. These matings may be so violent that if the male has managed to insert his arm into the female's siphon, it may be literally torn from his body. After such an encounter, the female can be seen swimming alone, bearing the grisly memento of a previous coupling. In time, the arm will be absorbed into her body.

But it is not always so rough. In the calmer matings, once the male has sidled up to the female, he gently inserts his copulatory arm into her, perhaps again and again, finally to leave it there. She normally doesn't object, and for the next several hours they lie quietly together *in copula*. The only sign of what they are doing is the deep heaving of the female. A series of sperm packets then moves slowly from a storage place under the male's bulbous head through a groove in his arm, finally to slip into the mantle cavity of the female. The sea water within her cavity ruptures the packet and releases the thrashing sperm; she then expels her eggs from her ovaries, the eggs and sperm unite, and the tiny fertilized ova are washed from her body into the sea, each to begin a lonely life on the ocean floor.

There is another strange aspect to the sex life of this intriguing animal. In some species, the copulatory arm of the male loads itself with sperm packets and then breaks away from his body to swim through the water as an "independent" animal. These wandering arms somehow are able to seek out females and enter their siphons, faithfully depositing the packets. Then the arms die, their corpses hanging from the female's body, producing the strange "nine-armed octopuses" which were the source of incessant speculation until the mating habits of the octopus became better known.

BEDBUGS

Traumatic Sex

It's probably not a good idea to complain of bedbugs. Being afflicted with ulcers is "in"; being plagued by bedbugs is not. The pesky insect hides in the cracks and crevices of mattresses, creeping forth at night to suck our blood. No one likes them, and they have apparently plagued us for eons. Their remains have been found in Stone Age caves as well as Egyptian tombs. Of course, enterprising humans have looked for a silver lining. It turns out that when bedbugs sniff human flesh, they give little "yips" of excitement. So U.S. Army scientists arranged to parachute them with tiny radio transmitters into enemy hideouts in southeast Asia to see whether the areas were occupied. The yips would be the giveaway. Even when performing a service, then, the animal seems less than honorable. The bedbug is a miserable creature indeed, so perhaps you won't be too sorry to hear how it reproduces.

First of all, even the most basic requirement for copulation is not met. The male bedbug does not put his penis into the female's hairy little vagina. Instead, he straddles the female's back obliquely so that he lies at an angle to her body midline. He holds her tightly, and, once firmly in position, he flexes his abdomen downward, pushing against her with his stiffened penis. The problem is, his penis doesn't reach her vagina. Instead, it reaches only a short way down her body so it is *here* that he penetrates her. The problem is that he *really* penetrates her, ejaculating his sperm into her body directly through her back!

The sharp point of his large, warped penis (which couldn't possibly fit into her vagina anyway) is forced through her tough body covering directly into a clump of tissue called the Organ of Berlese. When the organ has become packed with sperm, the sperm break out into the body cavity and migrate to special pouches where they will be stored for a while. Actually, they will stay there until the female has her next blood meal. That's you. When she has gorged herself and becomes distended with blood, the sperm become aroused and begin their migration to the ovaries where each egg will be fertilized as it is produced, the whole process having been set in motion by the nutrient blood.

The sperm's odyssey is a strange one. They do not migrate to the ovaries by taking the easy routes through the body's passages, but for some reason they move through the tissue itself. This gives rise to a host of puzzles. How do they know where to go? And when? And another thing—their path is now more perilous because the female's tissue harbors special predatory cells which try to trap and devour the sperm as they pass.

The female bedbug, of course, has adapted to this mode of copulation and has an array of internal devices to ensure that the sperm, in fact, reach the egg and that she survives the stabbing. So that she doesn't bleed to death she has a soft, fleshy organ lying near the site of the penetration which acts to block the wound and retard seepage. Also, she has developed the ability to store sperm. This may be crucial to her survival, because too frequent matings can spell her death. In fact, a female which mates with more than six males is likely to die from her wounds.

It has been suggested (under a variety of circumstances) that semen has a nutritive value. As far as bedbugs are concerned, as a matter of fact, the female's tissues do tend to capture and digest sperm on their trek to the ovaries. So perhaps insemination in these species is a way of getting food. The argument is bolstered by evidence that homosexuality is rampant among bedbugs. In fact, the males of some species have special organs, similar to those of females, with which they receive the sperm of other males. And some males show the scars of quite an active homosexual history. As an aside, an inseminated male bedbug may later ejaculate both his own sperm *and* those of his homosexual partner into a female, an occurrence which Darwinian evolutionists are indeed hard pressed to explain.

PIGS

Screwing

Copulation in pigs may be the source for the colloquial expression, "screwing," because of the peculiar design of the boar's penis. It is slender and surprisingly long—in fact, about a foot and a half in length. Also, the opening is not at the tip as one might expect, but displaced to the left of the shaft. However, it is the tip itself which is so unusual. It is not composed of erectile tissue like the rest of the penis, but of a dense, nonexpandable tissue which is spirally twisted into a kind of corkscrew.

The sow is not put off by the elaborate organ, however. In fact, her own plumbing is a bit unusual itself (as one might expect since the reproductive organs of the sexes would have to be coadapted). When she is young, she has a hymen, or "maidenhead," which is absent in old sows, and she has a very long clitoris ending in a small point which protrudes out of the vulva. The cervix (the opening between the vagina and the uterus) bears a number of tough ridges which are twisted so that they correspond to the male's own spirals. The opening into the uterus is also very small, just large enough to accommodate the tip of the boar's tapering penis.

When the female begins estrus or "heat," the vulva becomes swollen and reddened and begins to discharge large amounts of mucus. Her scent may seem a bit rank to the human nose, but a boar finds it irresistible. He becomes very excited, following her around, nosing, licking, and nudging at her genitals. When she stops moving, he rests his head

What you see here is but the tip of the iceberg. The boar's penis, actually about a foot and a half long, will drive the cockscrew-like tip far up, into the hard, twisted ridges of the female's uterus, where it will be held fast as it releases enormous amounts of sperm.

heavily on her back, quickly slides his penis out of its sheath, and mounts her. When his long penis is inserted far up into the mucus-filled vagina, he begins to thrust. The thrusting causes the penis, with its corkscrew tip, to rotate. And the longer he thrusts, the deeper goes his penis. Finally its tip reaches the spiral folds of her cervix, and, by rotating its way in, it becomes tightly lodged in the firm cervical folds. Once he is firmly in position, her long vagina begins to contract rhythmically against the shaft of his penis, bringing on his ejaculation. His ejaculation, by the way, is a bit spectacular itself. In other animals, small amounts of semen are normally ejaculated into the vagina from where they somehow thrash their way into the uterus. The boar, however, has taken the guesswork out by ejaculating up to a *pint* of semen directly into the uterus where his twisted penis is held fast. He also maximizes his success rate by ejaculating again and again, each time pumping enormous quantities of semen into the female's body! One would think all this would be quite enough to insure fertilization, but the boar has one more trick. The last part of the ejaculation has a tapioca-like consistency which enables it to back up and plug the entrance of the cervix, keeping the vast reservoir of sperm from leaking out.

FLEAS

Howe Fyne an Organe

A flea trap was at one time something other than a downtown hotel. It was once a trap for fleas—and it was usually worn by women, because women were more likely to harbor fleas than were men. Only recently have we learned that the flea's fascination with women may be strangely related to the female hormones. But let's start at the beginning, with the strange case of the European rabbit flea.

First let's see what the female flea is up against. The truth is, the male flea has the most complex penis known. It is actually composed of *two* rods, one thick and one thin, the ends of which slide forward as the coiled rods unwind like watch springs. The sperm themselves are wound in a mass around the end of the thinner rod, like spaghetti around a fork. The thin rod is longer and runs through a groove in the flattened end of the thicker rod, like a rope runs over a pulley. When the flea copulates, the thick rod is first pushed snugly into a blind pouch about halfway up the female's long, winding reproductive tract. This properly positions the sperm-bearing thinner rod so that it is able to continue on and probe into the tiny opening which marks the female's continuing vagina. The thin rod finally pushes the sperm into the sperm-storage organ of the female and somehow releases them there.

Some male fleas have an odd organ resembling a pussy willow twig or a feather duster with which they gently stroke the female while they copulate with her. It seems like a thoughtful gesture, but actually the delicate behavior is somewhat surprising since the penis is so loaded with spines, braces, and hooks that it can only be inserted by brute force. And, in fact, the female is often buffeted and torn quite badly during coitus, often sustaining serious injury.

One of the most fascinating aspects of sex in the European rabbit flea is that it is dependent on the sex-hormone cycles of its host. In fact, it can't breed unless it is residing on a pregnant rabbit.

The fleas begin to prepare themselves for reproduction as soon as the buck and doe rabbits set eyes on each other. The fleas know the fun is about to start by the sudden rise in the temperature of the amorous rabbits as each flushes in the sexual excitement of the moment. At this time the fleas also become excited, and when the rabbits approach each other they hop madly back and forth from one to the other. Finally, however, most of them settle on the doe and stay with her when the pair separates. Then, ten days before the young rabbits are to be born, the eggs of the female fleas begin to develop, triggered by circulating hormones in the mother rabbit's blood. The timing is such that when the rabbit is ready to give birth, so are the fleas. Soon after the young rabbits are born, the fleas move from the mother rabbit's ears (where they normally reside) to her face. Then, while she licks and nuzzles her newborn, the fleas jump to her young and begin to feed on their blood—mating and laying eggs all the while. After about a twelve-day orgy, they suddenly abandon the young rabbits and return to the mother to await her next pregnancy, but they leave their eggs to hatch on the young rabbits.

If other fleas are also dependent upon their hosts' hormone cycles, it may be that the women pictured in old catalogs wearing the latest flea traps were more likely to be harboring fleas, not because they were fairer or softer of skin than men, but because of their monthly surge in hormones.

BATS

Sleep, Fair Lady

It's actually pretty hard to catch bats copulating or doing much of anything else, because they're usually zooming around in the dark or lurking deep in caves. But even this furtive and elusive creature has been caught in the act by persevering, patient, and prurient observers. However, it must be admitted there are many kinds of bats, and even our small army of voyeurs have seen only a few species actually doing it.

We do know that most bats copulate in the fall or winter and that, strangely enough, there are no eggs in the female tract at this time. This means the female must harbor the sperm for months until her eggs develop in the spring. It also means that the bats must copulate at the very time they tend to be hibernating. And one must admit that it's a good trick to do it while hanging by your toes.

Apparently the males in the hibernating period are a bit more restless than the torpid females, and from time to time they rouse from their deep sleep, driven by a powerful reproductive urge, and grope toward the nearest female. Strangely enough, the testicles of the bat don't even become active until the breeding season. Only then do they produce sperm and sex hormones. And then, stranger yet, they begin to migrate. They move from deep in the abdominal cavity downward (upward?) to the scrotum, where they find it cooler and more amenable to sperm production. Once the sperm are produced, they may be stored for long periods in peculiar bulges in the tail membrane.

When the male has reached his full reproductive state, the surge in his hormones compels him to find a female. He clambers over the suspended bodies of his colleagues until some female attracts his attention. If she is awake, they may go through a brief courtship ceremony, nudging and stroking each other with their cold, velvet wing membranes. The male nuzzles her face, belly, and breast, but when the time comes he moves behind her to copulate.

If he has found her deep in hibernation, he omits the foreplay and simply moves into position behind her. In some species, he then grabs her by the nape of the neck, steadying her with his enveloping wings, while he pushes against her tail membrane with his pelvis until his erect penis reaches her rear.

The bat penis is an interesting organ in that it is curved and can move like a kind of lever, pivoting at its base. And it has another neat trick. It can move by itself, without the male making pelvic thrusts. Thrusting and jostling, of course, is not a good idea if one is hanging by one's toes above a hard floor. Better to let the penis do the driving.

The long organ, stiffened by an internal bone, begins by probing the female's rear, sometimes touching first the anus and the surrounding areas. But finally it discovers the vagina and slips in. Then, nothing. Both bats now lie motionless for perhaps minutes or even hours. Actually the male couldn't leave if he wanted to, since the tip of his penis is so swollen that it cannot be withdrawn. The copulation is actually an extended process. The male, held fast, continually ejaculates deep into the vagina over long periods of time.

Whereas the male moves scarcely at all, the female's abdomen pulses rhythmically. The heaving may be caused by rapid breathing; possibly the movements are muscular contractions of her vagina—a way to make the male ejaculate. They may also be her way of drawing the sperm into the upper reaches of her uterus.

In any case, after he has ejaculated, the male usually sim-

ply withdraws and finds his way to a resting place, but sometimes he doesn't quite make it. In his afterglow, drowsiness
may overcome him on the spot and he may succumb to the
irresistible call of that strange winter sleep. Thus the pair
spends the winter firmly locked in embrace.

One would think the embryo would develop in the normal
way of mammals. Why should it not? It turns out that if it
did, it would be born in the dead of winter. But the mother
has a solution. She simply almost stops its growth. This means
she can remain pregnant for many months instead of just a
few weeks. Thus her single infant will be born in the warm
springtime.

When the infant is born, it will ignore the two nipples at
her breast, except to feed, and will instead spend most of its
time clinging to two smaller "false tits" near her pelvic region. These yield no milk but serve as anchors so that the
young animal begins its life "heads up."

Once the mother leaves her infant and flies out to forage,
she may never see it again. The reason is, when the horde
returns to the cave at dusk, the bats settle randomly among
the young and the females nurse whichever infants find them
first.

CATS

A Hasty Withdrawal

Yowling cats in their mating rituals have undoubtedly aroused squint-eyed animosity in us all. Mainly, it's the males. In the spring they become particularly aggressive as each carves out his breeding territory. (This is important because some males are completely impotent outside their territories.) To begin with, each male carefuly marks his area with his very own distinctive mixture of urine and a kind of acid which comes from a special gland near his anus. The concoction produces that potent "catty" smell with which house cats like to mark the furniture. No matter how strong they smell, however, not all tomcats are successful in winning a territory. Some lose. If a tomcat is defeated in a territorial battle, he may become a nervous wreck, his confidence shattered, leaving him unable to breed at all.

The female also has her own scent in the breeding season, but fortunately it is almost imperceptible to humans. It is not imperceptible to a tomcat, however. In fact, the odor drives him mad with lust. Females he had previously ignored become the objects of his undivided attention. As is only seemly, the female's own behavior also begins to change. New hormones surge in her blood, and she begins to rub her head coyly against everything in sight, frequently crouching in a peculiar way, with her belly on the ground and her rear end held high, exposing the lips of her red and swollen vulva.

When the male approaches, the female elevates her rear even further and pulls her tail tightly to one side, treading

In many carnivorous species the male grasps the female's neck while copulating. Sex among cats can be very traumatic and dangerous. The male's penis is covered with backward-directed barbs and may be the cause of her spiteful behavior.

the ground anxiously with her hind feet. He moves quickly. The copulatory act itself is brief. He goes to her proffered rear and, wasting no time, immediately mounts by clasping her flank with his forelegs. He grasps her neck in his jaws and thrusts quickly several times until his penis, which points backwards and downwards and is stiffened by a small bone, has been pushed downward into her vagina. As soon as he enters her, he stops thrusting and becomes completely rigid as he quickly ejaculates.

Her clitoris, which also contains a long, tapering bone, probably is not greatly stimulated by such a brief encounter. Most of his thrusting, of course, took place before he penetrated her. One might wonder if his total unconcern with her pleasure has anything to do with her subsequent pique towards him.

Actually, he must be wary of her displeasure. After all, she is armed. So now comes the tricky part. She signals the end of the act by trying to roll and turn under the male. He must then dismount by *leaping* backward, landing well out of her reach, because when he withdraws his penis, she will invariably let out a bloodcurdling shriek and make a quick attack in his direction. Admittedly, her demeanor could be due to the fact that his penis is covered with backwardly directed horny barbs.

Once he is well away, she goes through a rather stereotyped pattern, rolling around on the ground and frantically licking at her vulva. For some reason, she is particularly dangerous at this time for tomcats and humans alike. She must be allowed to finish her ritual. But then, after resting for less than an hour, she is usually ready to go again, and, of course, so is he. They will repeat the entire pattern over and over for up to four days as the male, uninterested in anything else, becomes gaunt and wan from lack of food. She will give birth about two months later and then may come into heat again only two weeks after the kittens are weaned. She will normally give birth about twice a year for about nine years (although some cats have lived over thirty years, providing us with untold boxes of free kittens).

DOGS

All Tied Up

They remain a long time uncomfortable, not managing to free themselves until long after their desire has turned to disgust, a grotesque and lamentable symbol of many a human liaison.

—De Gourmont

Probably all of us have seen dogs doing it, whether we wanted to or not. They are bolder about it than cats are and less restricted than, say, cattle. So in the fall and spring, there they are—congregations of dogs attentively trotting after a bitch, quarreling and fighting among themselves while occasionally one attempts to mount her. And you may have been startled as a youth to see two dogs abashedly facing in opposite directions, their rear ends somehow firmly joined—at least until some morally outraged soul tossed a bucket of water on the sinners.

The frantic hubbub in a pack of dogs trailing an estrous female may veil the true male-female interaction—the one which occurs when a pair is alone, untroubled by squabbling competitors. However, it must be admitted that solitude does not ensure a normal mating. Some females simply refuse to mate with certain males while eagerly accepting others. (We have probably all had personal experience with this sort of phenomenon.) Should the male with her strike her fancy, however, she assumes an active role in the sex play, bounding playfully around him, biting at his neck and ears and attempting to get the idea across by mounting *him*. The male may be busy mouthing at the bitch and licking her vulva. His persistent licking arouses her, and her blood pressure rises sharply. Finally she stands with her tail pulled to one side, looking over her shoulder, her vulva strangely protruded while it performs a most peculiar invitation: rhythmically moving up and down. It's probably more to the point than moving one's eyebrows up and down, but the idea seems to be the same.

The male usually wastes little time in mounting her, holding her tightly around the haunches while he thrusts vigorously with his pelvis. His long, pointed penis extends from its sheath and is repeatedly shoved against her rear until it strikes home. Once the penis enters her, its tip is pushed far up into her vagina. Then, when his penis is firmly in place, its tip begins to flush with blood, causing it to swell. At the same time, muscles in her vagina begin to close tightly behind the bulb which is forming at the end of his penis, locking it firmly in place. For some reason, he then dismounts by throwing one rear leg over her back so that now he has all four feet on the ground and stands facing the other way. A good deal of traction is placed on the penis at this time, and this is usually when someone appears with the water.

So there they stand, his penis pulled backwards and both of them looking a bit ridiculous as they occasionally glance over their shoulders. The romance seems to be gone now, but there they must remain.

It has been suggested that the pattern evolved so that the pair could defend itself from threats coming from either direction, but this explanation seems somehow lacking.

The male ejaculates very slowly. The time can be measured by those interested in such things by counting the pulsations at the base of the penis. The semen steadily leaks into the uterus as long as the animals stand tied together. Anywhere from five minutes to an hour later the male is spent; the swollen tip returns to normal and the female's vagina relaxes its grip on the penis. They then go their separate ways. The male cannot be enticed again until he has rested.

PORCUPINES

Very Carefully

An old joke goes, "How do porcupines do it? Very carefully." And in this case our raunchy humor has not led us astray. They *do* do it very carefully. A porcupine's quills are only funny to those who haven't encountered them. They are long, bristling with backwardly directed spines, and once they enter the body they tend to swell, working their way deep into tissues, destroying organs and causing infection. The male has to be very careful indeed.

To make matters worse, female porcupines are bitchy, squalling beasts when they do not want to mate, and they may even swat at amorous males with their spiked tails. But when the mating season approaches, the female's behavior changes. Strangely enough, just at the time she develops an intense need for genital stimulation, her vulva grows more and more insensitive. Thus she spends more and more time stimulating herself by rubbing her rear on practically anything she can find. She especially appreciates a good stick which she can straddle. She is now very active and excitable, running about and gnawing on everything in sight. Then, at the period of greatest sexual desire, she becomes "morose," moping about, but still giving great attention to anything that touches her vagina.

If she encounters a male about this time, she may stand up on her hind legs and walk toward him, her belly exposed. Then she may drop to all fours and nose at his head, lightly touching the long, sensitive hairs on his sides. To make herself emphatically clear, she may stand, look directly at him, and "hump" the air.

The male porcupine has his own act—one that is a bit startling in its own right. He responds to a female in heat by nosing at her and making high, falsetto, whimpering noises. He then rears up on his hind legs and walks toward her, his unsheathed penis fully erect. As if this behavior weren't crass enough, when he gets within six or seven feet of her, he begins to drench her with short spurts of urine (a good trick in itself, since not many mammals can urinate with an erection, and six feet is not bad in any case). This unchivalrous soaking can be a bit much for her and she may amble away, perhaps to think through the entire matter one more time. Even if she stays, she doesn't seem to be too happy with him and may growl and snap and box at him with her paws until he stops this urinating business. If she is in full heat, however, she may be a bit more tolerant, even standing on her hind legs and allowing herself to be unceremoniously soaked with urine and a glandular discharge of his very own making.

He, of course, is happier if she stays. In fact, if she leaves he may be so frustrated that he picks up a stick and drags it between his legs so that it rubs against his penis and anal glands. He seems to prefer the same sticks she has used to stimulate herself. Sometimes one can find sticks in the woods which literally reek of the powerful essence of porcupine.

If the female does not flee from his advances, he may approach her, hobbling on three legs, clutching at his genitals with his left paw. By this time, she may be fully prepared to accept him and flattens her spines and pulls her tail to one side. She may also hold her tail straight up in the air while she walks backward toward him. This is the big moment. He meets her halfway by walking toward her on his hind legs. When they meet, he pushes his pelvis forward toward the only safe place—her genital area, safely covered with soft hair. Now comes the problem. If her tail is not properly positioned, he could injure himself on those particularly long, sharp spines which extend back from her sides. At this point, young male porcupines may be at a special disadvantage for two reasons: first, they have short penises which won't reach unless everything is done perfectly; and secondly, they don't know how to deal with the female's bristling tail if she fails to move it. Older males have longer penises, and as for the tail, they have learned just to push it out of the way.

Once the male has inserted his penis, he thrusts in a peculiar manner by flexing his knees and, stranger yet, by pushing against the ground with his tail (a most unusual use of the tail). The result is a kind of rocking motion. He may actually touch the female only with his genitals, holding his forepaws safely in the air as he rocks along. Sometimes, if he can manage it, he may touch her very lightly on the sides.

The female porcupine is not exactly passive through all this. When they begin, she lowers her head down to the ground in a determined manner and forcefully drives her rear against the thrusts of the male. The male quickly ejaculates, hesitates a second, and withdraws his penis. Then he sits back on his haunches and bends over, taking his penis in his mouth or licking at it frantically.

The pair will copulate again and again with twenty-minute rest periods until the male is simply unable to respond any more. The female, on the other hand, is not so easily sated. In fact, she is now completely aroused and more amorous than ever. But to no avail. The prickly old boar saunters away for a long rest—alone.

EARTHWORMS

A Little Give and Take

Hermaphroditus, you may or may not recall, was the beautiful son of Hermes and Aphrodite who was united with a water nymph at the Carian fountain and thus his body became both male and female. Earthworms may not be particularly beautiful, but they are hermaphroditic in that each worm has both male and female parts. (Thus one generally doesn't refer to a "boy worm.") Hermaphroditism, by its nature, might be considered vulgar or untoward in some circles, but it certainly solves problems for slow and sluggish creatures which get around so tediously that they might never encounter another worm of the "opposite" sex. As it is, they need only encounter another worm.

When two worms do find each other, they lie side by side, head to tail (except, perhaps, in those states in which such behavior is illegal). They do this because the testes lie toward the rear and the ovaries toward the head, so this position brings their gonads as close together as possible—but, it turns out, not into direct contact.

Even the segments of earthworms have not escaped being numbered; thus it can be determined that the fifteenth segment of each worm comes to lie along the large, fleshy ring, or clitellum (Latin for "saddle"), of its partner. The two worms then secrete a mucus covering around themselves and, while lying in their slimy pool, each drives tiny, hairlike spines into the body of the other. (Earthworms also use these tiny spikes to hold the ground when they are crawling.)

Earthworms lie head to tail and exchange sperm. The sperm are then stored in special receptacles. As a band containing the eggs slides down the worm, the sperm are released into the band, which then slides off the end of the worm. For some reason, usually only one offspring will survive.

Thus the mating worms become loosely bound to each other. Then sperm begins to flow from tiny openings in the fifteenth segments of both worms. The sperm of each moves in opposite directions, through a tiny groove along the bodies, a short distance to several tiny pockets near the head of the other worm (in the ninth segment). The sperm then flows into these storage pockets to await the next step.

In time, the two worms twist and writhe, pull away from each other, and go their separate ways. Later, after the eggs have matured, the peculiar clitellum, which lies back at about the thirty-third segment, begins to secrete a kind of nutritive slime band around the worm. The band then mysteriously begins to slide toward the worm's head. On its way to the segments harboring the stored sperm, it must first pass over the worm's ovaries. When it does, eggs are squeezed into it. The eggs then slide along with the band to be joined by the sperm which have been waiting quietly in their receptacles. When the band reaches the head, the worm writhes and contracts vigorously and shrugs off the slimy mass which then closes at each end so that no sperm or eggs escape. Only now are the eggs fertilized by the thrashing sperm. The slime quickly hardens, forming a crusty cocoon from which only a single baby earthworm will emerge. For some reason, only one embryo is "chosen." The bodies of its brothers and sisters are sacrificed and broken down for the nourishment of the fortunate one.

SHARKS

And Mermaids' Purses

Shark! The very word strikes terror even into the hearts of people sitting comfortably in Colorado. Perhaps this sort of respect for one of the most dim-witted of all vertebrates can explain why we know so little about them. When we see one, most of us clear the water.

Of course, not all sharks are dangerous. And this is another problem in describing their love lives. They differ from one species to the next. We have been able to observe the reproductive behavior of some small sharks and a few species in captivity, but we can really only guess at the sex life of the Great White.

In general, male sharks can be identified by two long, tapering appendages protruding from their undersurfaces. These "claspers" (a misnomer—they do not clasp) are stout rods of cartilage formed from modified fins. A clasper is not a solid structure but is "rolled" so that a deep groove is formed along its length. The tip, however, is expanded into a fanlike affair which scatters sperm in a radiating spray when the male ejaculates.

In some species the claspers are smooth, but in others they are covered with tiny hooks and barbs. These spines may help hold the penis in the vagina when a pair is copulating, or they may be used to rupture crudely the maidenheads of virgin females. In some species, the claspers are erectile—they swell with blood when the male becomes aroused. They are held tightly against the shark's body while he is swimming. But when he is ready to copulate, the twin appendages unfold and stand forward as he turns upside down and inserts one or both (usually one) into the vagina of the female. No matter how experienced she is, she always behaves like a nervous virgin, her body stiffening and remaining almost immobile through the whole thing.

In some species, the claspers seem to be used only to force the vagina open so that when he presses his body against hers, a fleshy mound which is the shark's true penis, arising from between the claspers, can enter the spread vagina. In yet other species, the sperm pours from his cloaca (the common opening for waste and sex cells in primitive animals) to run along the groove of the clasper, from where it is pumped by a rhythmically contracting organ at the base of the clasper directly into her body.

Species of sharks vary not only in their copulatory organs, but in their positions as well. Lemon sharks, for example, copulate side by side, their heads slightly apart but their bodies in such close contact that their swimming motions are completely coordinated. Since this position would require a very long penis, in this species, the sperm apparently runs along a groove in the clasper into the vagina. In the almost eellike common dogfish, the flexible male wraps himself completely around the female's waist, his body twisted so that his belly is pressed against hers, his claspers firmly inserted deep into her reproductive tract. In nurse sharks, on the other hand, the male grasps the hind edge of one of the female's pectoral (forward) fins in his jaws and forces her over onto her back so that he can slide one or both claspers into her vagina. In this species, males have large, prominent teeth, and so the more experienced females may bear tattered and scarred fins.

And, finally, in the horn shark, the male grasps the female by the left pectoral fin and swings his tail over her back as he braces himself against her second dorsal (back) fin. His body then forms a loop so that he can thrust his right clasper into her vagina. They may swim like this for over an hour as the male makes gentle, rhythmic movements with the hind portion of his body.

Later the females will lay either horny egg cases (called "mermaids' purses") which contain developing embryos, or will give birth to baby sharks. The mermaids' purses of sharks and rays are often found by beachcombers (although not many people know what they are). As an aside, in some species, embryonic sharks may not develop in eggs but may be nurtured by the mother's body in a manner similar to that of mammals. The mother secretes a "milk" from long filaments protruding from the uterine wall which the infant can then drink or absorb through its body wall. In some cases, the secretory filaments may extend, like long nipples, directly down the throat of the developing infant. Baby sharks are born ravenously hungry and may begin searching for food immediately. In fact, they are so voracious that, in some species, the first baby shark to develop in the uterus will turn and blindly devour all his brothers and sisters as they emerge from their egg cases inside their mother.

PLATYPUSES

Hell Hath No Fury Like a Woman Spurred

The platypus is a primitive mammal; it gives milk, but it is similar in many ways to snakes. It belongs to a group indelicately called the monotremes, which means "animals with one hole." This one hole is the cloaca, that common opening for solid and liquid wastes as well as reproductive cells—obviously a most uncivilized arrangement. If you were to peer into the cloaca of a male platypus (perhaps some Sunday afternoon when not much else was happening), you would see a bulge on the far wall of the chamber. That bulge is the platypus version of a penis.

Platypuses are difficult to observe, partly because they are shy, spending their time sifting small bottom-dwelling creatures from the sludge of muddy streams, and partly because their numbers have been decimated as trappers packed them off to zoos (where they usually promptly died). The observations we do have, however, tell us some strange things about their sex lives.

It all begins conventionally enough. First, of course, there is the wedding procession. It begins by the pair swimming in circles, the male firmly holding the female's tail in his bill. The ceremony is short, however, and without further ado, they jump right in and begin to copulate.

Platypuses are rare among mammals in that they copulate belly to belly, but they reject the "missionary position," opting for facing opposite directions, so that their tails lie across one another's belly. They do it under water so the details are a bit murky, but we know that they press the openings of their cloacas together. The male's erection lengthens his penis to about three inches until it protrudes from its cloaca, well into that of the female. This is the only function of the penis; urine leaves by a different route.

When the penis is inside the female, it blocks her urinary tubes so that the sperm have only one path open to them—up the two "sperm tubes" of the female to the waiting eggs. Since the female platypus has no vagina or uterus, the sperm must travel a considerable distance up these long, narrow passages. They are not left entirely to their own devices, though. They are assisted by the peculiar structure of the male's penis; it is split into two halves! When the pair is well into their act, each half of the penis actually enters a sperm tube of the female, where it immediately swells and is firmly held in place by the long spines which arise from each tip. Once both tips are in place, the male ejaculates. This part is a bit different, too. The sperm are ejected through a number of tiny holes at the end of each swollen tip.

The weird plumbing probably does not completely account for the female's general hesitance to have sex. The spines, of course, could be rather uncomfortable for her and could make her apprehensive, but in any case, she is hard to interest. The male, however, has a solution to this problem. He has two peculiar roosterlike spurs on his rear legs. The spurs are connected to poison sacs, but the weapons are almost never used in self-defense. Instead, it has been suggested that the male uses them on the female. The poison temporarily incapacitates her so that he can copulate with her with minimum bother. Since platypus sex is so peculiar to begin with, it should surprise no one to learn that the poison may also sexually arouse or intoxicate the female, thus increasing her willingness to mate. She has no clitoris, so perhaps this is her only means of sexual stimulation. Poisoning her, however, seems a bit drastic, since the toxin is a powerful blood coagulant which can actually kill small animals and incapacitate a man. The poison, however, is only secreted during the breeding season, a fact which lends credence to the mating theory.

In any case, a few weeks later the female will crawl into a hole she has dug in a riverbank and give birth, not to baby platypuses, but to two eggs! The tough, leathery eggs are small, about the size of a sparrow's, and will be incubated by the mother in a nest of sodden twigs and leaves. The baby platypus is only able to escape its egg by scraping its way out with a single horny "tooth" at the end of its beak, a tool which will fall off shortly after the baby emerges.

The mother has no pouch or nipples, so the young grope about, nudging her hairy underbelly and licking at the milk which oozes from the scattered milk glands. For the first few weeks, their bodies' thermostats are not very well developed, and they are as cold as any reptile. But later their temperature will level off at about eighty-eight degrees Fahrenheit, a cool temperature for a mammal and one that, with the eggs, belies the platypuses' ancient reptilian heritage.

OPOSSUMS

Split-Penis Possum

The opossum is one of evolution's successes. It, like the shark and horseshoe crab, has managed to survive almost unchanged in a wildly fluctuating world. And the opossum has managed to survive on land, where the changes have been most drastic. It is a member of that lowly group of mammals called the marsupials in which the females protect the young in pouches, at least for the early part of the infants' lives.

The male opossum may be like someone you know, ugly as sin but ready to have sex at the drop of a hat. The problem is, the females will only cooperate at certain times. The males can't complain too much, however, because the females are fairly active themselves, coming into heat first in January and bearing perhaps two or three litters each year. It is easy to tell when the opossums are going through a reproductive period, because the male prowls around making peculiar metallic clicking noises. No one knows how he does it, and he's not telling. Then, he has another peculiar habit. He licks things and then rubs his head on the wet place, leaving his scent everywhere and making his hair look as if he's on his first date.

When copulating, the male grabs the female by the nape of the neck. A lot of animals do that so it's not so surprising, but his behavior after that becomes somewhat indelicate. He manages to wrap *both* his front and back legs around the female before bringing his penis down against her vagina. In the opossum, the testicles hang in a long, pendulous sack in *front* of the penis. So down is the way to go. Once he has entered her he falls over, so that they lie on their right sides. For the next twenty or thirty minutes, they will lie there firmly joined, the male sporadically thrusting, a grimace on his wide mouth.

Possum plumbing is rather unusual among mammals in that the vagina and uterus of the female are divided, essentially forming two organs, at least at their uppermost parts. Since she has duplicated some of her parts, the male has duplicated some of his. His penis has divided into two parts. When the penis slides out of the foreskin, its peculiar forked arrangement becomes immediately apparent. Thus when he copulates, he is able to service both forks of the vagina at once.

Actually, in the opossum, the male urethra (the tube carrying both urine and sperm) is not really a tube. It is a groove on the inside of each half of the penis. When the halves are together, then, a sort of tube is formed. Thus a careful observer in the proper position will note that the male opossum can urinate in one stream—a fascinating bit of trivia. In copulation, since the halves are separated, the sperm must travel along the grooves toward the pointed ends of both halves of the penis. These grooves become more shallow and finally disappear short of the end of the penis. Thus the sperm runs out all over the place.

It all seems, somehow, that the system would not be very efficient. It is efficient enough, however, for the female to give birth to up to twenty-two young a few weeks later. For the opossum, birth is a relatively painless process. The tiny young emerge at a very early stage of development, and they assist their mother's labor by half crawling out of her vagina. Once they are out, they unerringly squirm their way upward over the vulva where their mother licks them clean of their birth fluids, and then they wend their way to the protection of the pouch using a peculiar overhand motion similar to the Australian crawl. When the mother starts to give birth, she sits down, feet outstretched before her, thus bringing her vagina closer to the pouch. She has prepared everything for their arrival. By the time they make it to the pouch, she has already cleaned it out, not once, but several times.

No matter how many are born, probably only eleven will survive, because that's how many nipples there are. Each undeveloped "larva" will attach firmly to a nipple, its own flesh somehow fusing to that of the mother. And there it will remain for about the next fifty days, growing swiftly all the while. At the end of that time, several hairy little gremlins will emerge from their secure world to begin a somewhat more precarious existence, clinging to their mother's back as she makes her nightly rounds of persimmon trees and garbage cans.

SEAHORSES

Pregnant Fathers

In the animal world, if anyone gets stuck with the kids, it's almost always the mother. In only a few species is the father left to tend the eggs or young, but the seahorse has carried this trend to its extreme.

There is nothing in the seahorse's courtship to suggest that the male is about to be left holding the bag. Not at first. The courtship is an elaborate affair, with the two animals approaching each other, dragging their tails on the sea bed, the male with his head petulantly tucked against his breast. Of course, in the courtship ceremony of most animals, the male tries to persuade a female to accept his sperm. In the seahorse, though, the female coerces the male into accepting her tiny, unfertilized eggs. She swims around the male in her resplendent colors and finally grasps him with her prehensile tail. They rock to and fro until, with a quick toss of her head, she indicates that it is time to head toward the surface. As they rise, the male puffs up his abdominal sac. She then begins to probe his body with her "penis," actually a nipple of flesh, until she finds the opening to his pouch. As they swim belly to belly, she pushes her bright orange "penis" into his pouch and ejects up to 600 eggs into it. Then, with no further responsibility, she withdraws and swims away, free as a bird.

The male's own genital tract opens directly into his pouch, and the presence of the eggs has induced him to release sperm over them. His pouch is lined with specialized nutritive tissue

47

Among seahorses, the female deposits eggs into the male's pouch with her tiny "penis." This triggers the release of sperm by the male. He will then be pregnant until he ejects a horde of tiny seahorses by violent abdominal contractions.

across which food and oxygen can pass to nourish the eggs. In other words, his pouch acts very much as a uterus! The eggs will develop and hatch inside the pouch until his belly is swollen with hordes of perfectly shaped, tiny seahorses. About fifty days later, when his "pregnancy" is terminated, he will begin labor, with spasmodic contractions contorting his entire body again and again. With each painful contraction, one or more tiny seahorses are expelled until the water is aswarm with little bodies darting in confusion hither and yon. The male will continue his vigorous heaving and wrenching until the last infant is expelled. It is important that he does. Should he fail to completely empty his pouch, the decaying corpse of even a single offspring can spell his death.

WHALES

Venus Rising

Whales are enormous creatures, and they do it in the ocean. So we don't have many good accounts of their love lives. No diver with all his faculties would want to be too close to copulating whales. From the observations we do have, though, we have learned that each species does it its own way.

The huge blue whales are among the most circumspect copulators. They court partly in the briny depths, away from all prying eyes. On the other hand, the humpbacks seem not to give a hoot. They are lusty and exuberant creatures. In their foreplay, the males may ecstatically hurl their hundred-thousand-pound-hulks clear of the water again and again before they settle under the dark surface to mate. Whalers call the female humpbacks the "whores of the sea" because of their willingness to accept any and every male in the ocean—first come, first served. But even whalers have wondered, how do they actually *do* it?

The blues and humpbacks are large whales, and so we should expect their sexual organs to be large also. But one's first view of a whale's penis can be an unforgettable experience. It may be blunt or tapering, and when it is retracted, it is coiled by powerful muscles deep in the abdominal wall. The tip is sheathed in a fleshy fold under the belly, some distance from the anus, and when retracted, the whole thing is rather inconspicuous. But a whale with an erection is a sight to behold. The brightly colored and elastic penis now stoutly stands forth, a foot in diameter and a full ten feet long! The testicles, by the way, are impressive in their own right, lying deep within the body and producing enormous amounts of sperm.

So how has the female accommodated to such an organ? Her vulva is a long, elliptical groove just above the anus. During sex it flares open, and the enormous clitoris nestled inside is clearly visible. On either side of the vulva are the two nipples which will nourish the seven-foot offspring born about a year (a surprisingly short time) after intercourse. A virgin whale has a hymen—or maidenhead—a thick strand of tissue about five inches long and a half inch in diameter, which stretches across the star-shaped opening of the vagina and which fails to survive her first intercourse. When a female, young or old, comes into heat, the vulva secretes copious amounts of slippery mucus. Whales may copulate at any time of the year, but it is most common as they leave the food-laden polar waters to begin their annual swim toward the warmer temperate regions, and, as mentioned, the different species use a variety of techniques.

The gray whales copulate without great fanfare; they simply lie, belly to belly, near the surface. The whole thing is very civilized. There is no struggling or thrusting, and each episode lasts only about thirty seconds. Among right whales, the male simply achieves an erection and then swims upside down beneath the female. He arches his back once, and that's it. But happily, some whales do it with a bit more pizzazz.

Take the blue whales, for example. After a considerable period of foreplay, which involves gentle rubbing with the flippers, biting and mouthing, rolling over and around each other and nuzzling each other's genitals, the blue whales suddenly break away from each other and "sound," diving deep into the briny sea. Even in the utter blackness of that other world, they are in constant communication through their whines, whistles, and moans, even though they may by now be some distance apart. But at some signal, they turn toward the surface, and with their tails whipping mightily they rush together, surging at ever greater speed to break free of the surface just as the great penis of the male thrusts deep into the vagina of his paramour. In that instant, he ejaculates. They almost seem to hover for a second, their great bodies towering out of the water, before they fall with a thunderous splash back into the sea, and it is over. But after a little more nuzzling and courting, they may turn and perform the whole thing again. And then again.

Sailors, for years, described strange pounding crashes on still nights of the open ocean, never guessing the source of the noises as the great whales repeated their lusty saga again and again, whistling and chortling to each other and scattering lesser creatures before them.

SALAMANDERS

I Never Even Touched Her!

Presumably, we're all aware that the object of "doing it" is to bring eggs and sperm together. This usually means the male must get his sperm into the female's body. In the strange case of the salamander, this is what happens, all right. But technically she manages to remain a virgin! It's a good trick, so let's see how the aquatic salamanders accomplish it.

The males are quite garish and brightly colored, and their flattened tails, useful in swimming, can also present visual broadsides which females find irresistible. The males, not surprisingly, initiate the courtship. They follow the females by incessantly sniffing at them. The sniffing seems to annoy the female at first, and she moves away. But he follows—a pest, bumping, nudging, and sniffing, continually sniffing. Finally, though, when he figures he's wasted enough time with this sniffing business, he moves in front of her, crossing her path and cutting off her retreat. When she stops, he begins undulating his beautiful tail so as to show off the vivid markings, and at the same time, he sets up a current of water which bathes her in his own scent. Females being females, they often immediately attempt to leave the scene. But the male follows and cuts her off again. Gradually, if she is at all receptive, her behavior changes. She is a bit slower to leave and seems to find his tail more interesting. He continues to fan. Then at last she takes her first cautious steps towards him. As she draws nearer, he takes a few sidesteps and draws away, fanning all the while. If she continues to follow him,

53

Some salamanders do not go through the ritualized mating rite. In this species the male newt grasps the female in a stranglehold and hangs on until she lays her eggs, when he releases his sperm over them.

his behavior suddenly changes and he lowers his head, turns, and walks directly away from her in a peculiar stiff-legged fashion. She is now following close behind. So close, in fact, that she may bump his tail with her nose. This is the cue he has been waiting for. He now folds his tail, accordionlike, and suddenly raises it high, displaying two bold white stripes across his own cloaca. She probably doesn't see the small jellylike mass which slips out of his body and quickly sinks to the bottom to stand like a tiny toadstool, glued by its gelatinous stump. The cap of this toadstool, however, is aswarm with sperm. The female is oblivious to all this, however. Her attention is still focused on the male.

He then moves exactly one body length away from the

sperm mass and stops again. As she follows him, her body, already swollen with eggs, passes over the clump of sperm. His measurement was precise. When her nose bumps his side, her cloaca—which had everted, turning inside out when she took those first timorous steps—is positioned directly over the sperm. She can hardly keep from brushing it with her cloaca, but that is enough. The clump adheres to her opening (no one knows how), and as she moves away, the sperm cap, stuck fast, is gradually drawn up into her body, leaving only the gelatinous stalk sticking out. She never seems quite to grasp what has happened to her. The male has now gone, and she can be seen scraping her rear on the gravelly bottom, trying to rid herself of that peculiar soft blob protruding from her cloaca.

ELEPHANTS

Brace Yourself!

There is an animal called an elephant. . . . No larger can be found. They possess vast intelligence and memory. And they copulate back to back.

—12th Century Latin Bestiary

Since elephants are the largest land animals, we might well find ourselves curious about how they do it. Does size make any difference? Actually, it turns out that they do have to make certain accommodations to their great bulk, but they also have a number of peculiar sexual patterns which don't have anything to do with size.

For example, what is musth? From time to time, male African elephants secrete a dark, heavy substance from glands near their ears. The thick brown fluid streaks their faces as it runs down to the corners of their mouths. During musth, the great beasts become extremely dangerous, and even "tame" bulls must be chained down. No one can explain musth, but several theories have been offered. For example, it has been suggested that the increased aggressiveness of males during musth may be a means of permitting them to deal with the naturally ornery females so that mating can occur. Also, other males seem to fear and avoid a comrade in musth, so competition would be reduced. Musth, then, may well have something to do with mating. The problem is, elephants will mate whether they are in musth or not.

Wild elephants may or may not go through a brief court-

Luckily, the male elephant does not have to thrust against the female. He has remarkable control over his long penis, which probes her body as they both stand motionless.

ship ceremony. Usually, however, a pair just gradually begins to take more notice of each other as the female slowly comes into heat. As her receptivity increases, they may become more and more restless, finally forsaking all else, even eating. The male is particularly interested in the urine and dung of the female and frequently sniffs her vagina. He also spends a lot of time sniffing her toenails, for some reason. Who are we to say what is sexy for elephants?

The male also produces his own reproductive scent. Should he be standing upwind of a female when he unsheaths his penis in an erection, she doesn't even have to see it. At her first whiff, she will give a great sigh and immediately wheel and walk towards him. She will stand near him for long periods sniffing his penis, his mouth, and his temporal gland and may also take an inordinate interest in his feces and urine.

Finally they begin to interact a bit more directly as they intertwine their trunks and shake their great heads from side to side. At such time, they may open their mouths and push them together in a kind of very sloppy kiss. They're pretty excited by now and may tear leafy limbs from trees and shake them to and fro in their anticipation.

Eventually the male moves to the female's side and lays his trunk over her back. This always brings on an erection as his great penis, about four or five feet long, is pulled out of its sheath by a pair of powerful muscles. The penis at full erection has a peculiar S-shape, and the end swells into an enormous bulb.

The female had now better brace herself because the male's next step is to rear onto his hind legs and place his forelegs on her back. He supports much of his own weight on his hind legs, but she may have to stabilize herself by pushing her head into a riverbank or by curling her forelegs under her while her hind legs remain straight. Her urogenital orifice is now wide open, exposing a very large clitoris.

The male needs a long penis because the female's vagina is situated well up under her belly. In fact, the wrinkled vulva looks so much like a penis that sexes are hard to tell apart in the wild. Because of the peculiar location of the vulva, the male has to "hook" his penis into it. So although the penis is long, probably only a few inches ever actually enter the female. The uterus, however, may be *yards* away from the vulva, so the male has had to develop a high-pressure propulsion system to propel the semen far upward with each spurt. In addition, the bull must ejaculate enormous quantities to insure that some of the semen reaches its distant goal.

The female is fortunate in that once the male mounts her, he does not have to shift his great weight around in order to position himself properly. Instead, the penis does the positioning. It moves, snakelike, up and down and from side to side, searching for the vaginal opening. It's actually not too good at this (perhaps eyes would help). It often finds the anus by mistake, and the bull happily but uselessly ejaculates. When his search has been more successful, however, his penis vigorously jerks up and down, probing at the vagina again and again until it enters her. After its shallow penetration, there is a brief pause followed by the explosive ejaculation. They may repeat the act several times a day for a few days until the female is no longer receptive. Then she, perhaps herself not fully grown, will be pregnant for the next two years.

BEES

It Takes a Lot of Guts

Honeybees have attracted our attention from the time we were barefoot youngsters, trying to avoid stepping on them, until later when we learned that they can communicate by dancing. Actually, it seems, the more one learns about bees, the more fascinating they become. But particularly fascinating is that great sexist sisterhood of the hive.

As we know, each hive has its queen. She doesn't seem to be very different from—only a bit longer than—the other bees, but she is the only bee in the hive able to lay eggs. Furthermore, she is able to determine how many of her offspring will be males and how many females. Surprisingly, the queen is genetically no different from her lowly sisters, the workers. They all started life the same way, except that the queen's chamber was slightly larger than the cells of her sisters. Also, she was fed a special secretion from the facial glands of the "nurse bees." Actually, all larvae are fed the secretion, but the developing queen is fed it longer than are her sisters.

A queen will tolerate no usurpers. When a royal lady emerges, she will immediately seek out any other developing queens and murder them in their chambers. Should any escape long enough to develop into an adult, the two will fight to the death. There can be only one queen.

The victorious queen will then leave the hive and circle the area on a reconnaissance flight. When she returns, she will stay only a short while before she launches herself again. This time she flies fast and hard, and she is trailed by a swarm of drones, the hive's males. Depending on the species, she may mate only once in her lifetime—with the first male to catch her in midair. When a drone overtakes her, he grasps her with his powerful armored legs and, curving his belly under her, erects his penis hydrostatically by powerful muscular contractions and thrusts it, with its backwardly directed barbs, deep into her vagina. This will be his last act since he pays the ultimate price for his conquest. When he pulls away, his penis remains firmly embedded in her body, and he is disemboweled. He falls then, trailing his guts and organs of his lower abdomen, and is dead by the time he reaches the earth. But it is no matter; his role was to leave his sperm. She flies on, a grisly reminder hanging from her reproductive tract, but a trophy which plugs her vagina so that none of the precious sperm are lost.

The queen then pumps the sperm into special pouches deep within her body where they will be stored. This is bad news for the other drones, because it means they are now useless. Useless individuals are expendable because the hive must not be hindered in any way. Deadheads cannot be tolerated. So once the bloody queen returns, the hapless drones which follow her home are attacked by their sisters. Their antennae, wings, and legs are torn from their bodies as the merciless workers move impassively and persistently against them. Their abdomens are sawed from their bodies and lie strewn about, twitching uselessly on the hive floor. A powerful but helpless drone may drag a mass of relentless workers around in wide circles until finally, exhausted, he is flipped onto his back and dispatched by the army of "wrathful virgins." A few drones may find some nook or cranny to crawl into, but a constant guard of workers stationed before the entrance to their hideaway ensures their starvation. Some drones may manage to get away, but those few which escape know only one refuge when night falls. They go home. And each morning a detachment of workers clears the area of their brothers' corpses.

The queen may lay a few thousand eggs a day for perhaps as long as five years. If, when she lays an egg, she squeezes her sac of sperm, a worker—a sterile female—will be born. If she fails to exude a tiny drop of sperm-laden fluid, an egg with only half the number of chromosomes will be laid (they will lack those contributed by the father), and these eggs will develop into drones. The drones, then, have no father; only their mother, the queen. Eventually the queen may run out of sperm and can therefore produce only idle drones. This marks the end of her usefulness to the hive unless she can be reimpregnated.

SNAILS

Love's Arrow

As a group, snails must have the most bizarre sex lives of any animal. For example, they murder their own sperm, they undergo sex change, they practice group sex, and their love-making can kill. But let us not dally with theatrics. Let's consider two kinds of animals—a water snail and a land snail.

It was once thought that freshwater snails placed in the same tank would seek each other out in order to copulate. What really happens, though, is that they simply begin to circle—almost always counterclockwise—and unless one is slower or takes shortcuts, they may never meet. When one does overtake the other, though, it gets to be the male.

These snails have both male and female parts and so can play the role of either sex, or both. The current "male" climbs up on the shell of the "female" and slides forward to cover the protruded part of her body with its own broad, fleshy "foot." His sexual excitement shows in the peculiar way his body arches. His penis, located just behind the right tentacle (the fleshy stalk on the head), has previously been recognizable only as a whitish blur beneath the skin, but now it blanches whiter and grows into a prominent knob. From the knob the penis sheath itself slowly begins to emerge, just as one would slowly turn the finger of a glove inside out. The penis is enormous. It's rather flattened and about as long as the snail, tapered at either end, but the partner as yet has no hint of its great size. Perhaps this is an adaptation to keep her from changing her mind. The flattened edge (hood) of

Snails have incredibly bizarre sex lives that may include maiming, killing, spermicide, and sex changes. These land snails have just penetrated each other with sharpened love darts.

the male's body now gently gropes about the body of the female as the penis sheath impatiently swells and contracts. Finally, when the tip of the penis touches the vagina, a small whitish ring which lies well back near the shell, the penis suddenly springs out to its full length, with the tip inserted into the female's tract. It is not until the penis begins to enter the vagina that she seems to figure out what's happening. Whereas she had been ignoring the male up to this point, she suddenly contracts so violently that she may actually dislodge him. Once the penis is inserted, however, she again begins to ignore her suitor and may casually resume feeding. Although the male immediately ejaculates with a shudder, the couple may remain joined for anywhere from a few minutes to several hours, showing, perhaps, a certain sensitivity on his part. He is not necessarily a great lover, though, because sometimes he misses the vagina entirely and ejaculates against her body wall. No matter; he contentedly moves away, apparently quite satisfied with himself. One wonders if his great "sensitivity" coupled with lack of technique could be because he spends about half his time as a female. In fact, the next time they meet, the sex roles are likely to be reversed.

In some cases, something of an orgy may develop as the male, already astride a female, is approached by another snail and is himself mounted and (we don't have a socially acceptable single word to use here) "copulated with." So while his penis is inserted into the snail beneath him, another snail crawls aboard and violates "his" vagina. And then along comes another snail, so guess what happens to the one on top. Thus a chain can be built with the first individual acting solely as a female, the last solely as a male, and all the rest as both sexes at once.

Apparently these snails can fertilize themselves also, but no one knows how. These virgins have more trouble giving birth since their vaginas have not previously been enlarged by a penis.

We find another weird system among the aquatic slipper snails. In their youth they are all males, and as males, they tend to roam about. As they mature, however, they settle down in one place, attach to something, and become females. Actually, this "something" may be a full-fledged female, in which case the newcomer quickly copulates with her before his masculinity fades. Then a third male may come around and take his place atop the second animal, now a female. This process can be continued until the tower is composed of up to fourteen snails. Only the top individual is a male and his huge penis can be seen working away on the female below.

Now the land snail, the *escargot* of French cuisine, is unusual in that it has its sex organs in its head. And another thing—these snails *exchange* sperm, thus fertilizing each other at the same time. This is not so unusual among lower animals, but the startling violence of the sex act is. It all starts off peacefully enough. As two animals draw near to each other, they first gently reach out and touch each other's soft bodies, lightly probing here and there with their sensory stalks, rearing up, "kissing," and generally getting the feel of each other. Their slow movements and gentle touch give no warning of what happens next. As if on cue, each animal suddenly thrusts a sharp, chalky dart, up to an inch and a half long in some species, deep into the body of the other. The sculptured dart has flattened, cutting edges like those of the deer hunter's arrowhead, and as it is driven home, it causes the other animal to twitch with pain. The dart may imbed in the lung, gut, brain, or heart of the mate, killing it immediately. But if both should survive this strange form of stimulation, they will first draw apart, seemingly excited by the pain, and then slowly move together again. This time each immediately softly slips its astonishingly large penis deep into the genital opening of the other. Each snail ejaculates (apparently with an orgasm) as soon as its penis penetrates far enough to reach a bladderlike receptacle where the sperm will be held until it is time to fertilize the eggs. When the act is completed, the wounded snails fall away from each other and lie motionless for a time, as if stunned. Then they slowly recover and crawl away in separate directions.

Some time later, the sperm will move back down the reproductive tract to a place where it forks, this time taking a new route, one which leads to the eggs. By this time, the snail has managed to poison its own sperm so that any reaching its eggs will be those from another animal. The continual mixing of genes from different animals, after all, is what sex is all about.

PRAYING MANTISES

Ravishing, Isn't She?

The French biologist Jean Henri Fabré had this to say about how mantises do it: "I have seen it done with my own eyes, and have not recovered from my astonishment." Why? What was it that distressed Fabré? Perhaps a headless lover.

The praying mantis is an unusually powerful and very visual animal with eyes that are specialized for detecting movement. It has lightning-quick reflexes and a voracious appetite. It has been rightly called a "terrible predatory machine." The spindly insects have even been reported to devour frogs and overpower small birds. But what's this about a headless lover?

Let's retrace their mating ritual. The male is often smaller than the female, and once he has spotted her, he must approach with the greatest care. As she stands motionless, gazing fixedly in all directions with blank eyes, he may rest assured that he is in her field of vision. If she moves in any way, he instantly freezes, even if in midstep. To move would mean death. He must now wait until she has captured some insect or is otherwise occupied. He then resumes his approach, taking only a few careful steps at a time, each leg shifting very slowly; after each step he freezes. He may remain motionless for seconds, or a minute, before taking those next measured steps which draw him even closer to her powerful spiked arms. Finally he is literally within her striking range. He must act now. One might wonder why he is willing to risk his life for sex. Undoubtedly, "why" never

This male is lucky—so far. The large, powerful female may have already devoured several other suitors. Strangely enough, they become more amorous after she has eaten their heads.

crosses his mind. He is the descendant of generations of males which did take the chance. More circumspect males of each generation do not leave their "circumspect genes." As a descendant of sexually driven males, he is irresistibly drawn to her.

Now he makes his rush, half running and half flying. If he is quick and lucky, he will leap onto her back and her only move will be to recover from being jostled. Once astride her, he fits his forelegs into small grooves at the base of her wings and grips her tightly. His long abdomen then probes downward and to the left, directly toward the tip of her abdomen. After some probing, the two plates over her vagina separate (the first sign that she is aware of his existence). He finds the opening and immediately inserts the strange barbed contraption which is his penis. His abdomen now contracts spasmodically, pumping gelatinous flasks of sperm into her body which are ruptured by four small "teeth" near the opening of her vagina. The released sperm are then stored in a small pouch until she is darned ready to be pregnant.

After a while, the male spreads his wings, sits motionlessly for a moment, then with a crackling flutter lifts himself and sails to safety. If they have performed the act while hanging upside down from a twig (perhaps for variety), the male suspended from the female's back, he will first let go with his forelimbs and hang from her with small hooks on his rear legs. Then he lets go all at once, falling to the ground and scurrying away. She doesn't seem to appreciate his conquest, and after all his trouble, she may curl her abdomen under her and devour any sac of sperm still protruding from her vagina. Hopefully, some sperm will already have escaped into her body.

But this whole episode may have a more shocking ending. Some males make critical errors. As a male approaches, should he cross the midline of her fixed gaze, her folded arms will flash outward and snatch him to her breast so that he is held upside down under her. For his part, however, his task remains unchanged. He must mate with her. Before he can move, however, she begins to devour him. She starts with his eyes, then his head as he lies clasped to her bosom. But as she eats through his head, she destroys a certain clump of nerves which lies just beneath his throat. In this clump are the very nerves which have previously inhibited his sexual behavior so that it would not be discharged at the wrong time. Thus she has destroyed all his sexual inhibitions and the headless corpse begins to writhe and kick with his legs until he has rotated his body enough to reach her vagina with his copulatory organ. The penis is inserted, and the abdomen begins to pump with more vigor than was ever possible in life. As she continues to eat him, working her way downward, the vestige of his abdomen pumps sperm more and more frenziedly from its terminal end. The surging of the cadaver against his mate only stops when she finally reaches back to eat the last remnants of the soft twitching flesh which had housed his penis.

RABBITS

Want to See My Scut?

Is it true? Do rabbits do it, well, like rabbits? The rabbit, of course, is generally believed to be the personification of promiscuity, even having become a "sexy" symbol in certain rather leering circles. And whereas its reputation is not without some merit, the rabbit's libido, although a bit kinky, may be slightly overrated.

Few of us have ever seen wild rabbits doing it, but the scattered reports are indeed fascinating. What happens is, a group of male rabbits closely follow a female, each frantically trying to mount her. If she is not amenable to the idea, she may simply squat unhelpfully or run away. But if she is receptive to a male, she may show him her scut. In turn, he shows her his. The scut is the white undersides which are exposed when the hind legs are straightened a bit and the tail is held tightly against the back. At this time, a courting male may pay the lady the ultimate compliment by urinating on her and imbuing her with his very own scent—a most romantic gesture among rabbits.

She is overcome! Her back now arches high, and she stands on the very tips of her toes. The courting buck mounts her from behind, places his chin on her back, and rapidly flutters his pelvis against her. At this point, he may run into trouble if there are other bucks around with the same intention. The jealous suitors spend most of their time rushing furiously at each other, biting and kicking with malice toward all. So should a colleague actually manage to mount the female, he can expect to be vigorously pummeled. Sometimes a buck may leap over a courting couple and with his hind leg administer a sound drubbing to the head of his competitor. In other cases, all the bucks may pile on, each frantically mounting the other in a rather spectacular, if unproductive, orgy.

Should a male be successful, his backwardly directed penis will spring forward, and he will thrust eight or ten times until it enters her vagina (which, during her receptive period, is purple and filled with mucus). Her clitoris is well defined and, in fact, is about as large as his penis. Also, it is arranged in such a way that it is vigorously stimulated during copulation. When he enters her, he immediately ejaculates with a tremendous thrust that may send him sailing off her back. His withdrawal seems to be quite painful for some reason, and one or the other may let loose a piercing shriek. Any unpleasantness is apparently quickly forgotten, however, and they may have another go within a minute.

His sperm remains in her vagina because it was ejaculated far up near her cervix and because it was mixed with a thick viscous discharge which temporarily blocks the opening. Actually, the sperm of many males may be mixed, since the female often accepts a steady parade of suitors. In some cases, however, when one male finishes, she may run and crouch in the grass. For some reason, when she does this, the other males do not approach her for a while, chivalrously allowing her to rest. This is a good time to fight, though, so the bucks spend their time making quick, threatening rushes at each other.

Rabbits are one of only a handful of mammals in which the females do not release their eggs from their ovaries until they have copulated. If you're interested in the biology of all this, the vigorous action of the sex act stimulates the brain to secrete a hormone which causes the release of the eggs. No one knows just how it works, but the advantage is obvious. The female does not waste energy in forming eggs which will not be fertilized. The sperm can wait for the eggs for up to thirty hours in her rather hospitable uterus. There are actually two long, tubelike uteri, each opening into the vagina separately. The eggs are released about ten hours after copulation and quickly move down toward the waiting sperm.

In only a month, the female will give birth to up to five young. If she has only one or two infants, she may come into heat again even while nursing. But if she is nursing three or four babies, any embryos which are produced from the second mating will abort in their early stages. Thus the demands made on her are kept low. In some instances, she may copulate surprisingly soon after giving birth. In one case, a male was observed mounting a female who was actually engaged in the act of giving birth. She assumed the copulatory position with her newborn, just minutes old, literally hanging from her nipples. The male was only prevented by yet another offspring emerging from her bloody vulva. As she turned to eat the telltale afterbirth which might attract predators, the male quickly moved in and copulated with her on the spot.

ALLIGATORS

Side by Side

Some of the earliest reports of alligators and crocodiles doing it may not be entirely accurate. Nineteenth-century explorers described females being thrown on their backs by the males who then mounted them belly to belly and, after taking advantage of them, chivalrously righted the females before departing. Other witnesses stated darkly that they had seen males who had just overturned females chased away by the "most degenerate of men" who then availed themselves of the ladies' pleasures. Such behavior is undoubtedly in violation of certain local statutes, so these reports have been largely discounted.

The notion of belly-to-belly copulation probably stems from our not being able to see how else they could do it. It seems clear that the male *should* be on top. After all, alligators and crocodiles have cloacas, the common reproductive and excretory opening characteristic of amphibians, reptiles, and birds. The penis is erectile and fleshy (similar to a man's), but the sperm duct is not a tube. Instead it is an open groove. Therefore, if the male were on top, gravity could facilitate sperm flow. It all sounds logical enough, but it turns out that the male is actually *not* on top of the situation.

Part of our problem with understanding how these prehistoric relics do it is that almost no one has caught them in the act. In our southern swamps, the larger alligators have all but vanished as their hides appeared in the wardrobes of people with lots of class—albeit low. And African crocodiles

Courtship in alligators is surprisingly tender and some of it remains puzzling. Copulation is accomplished by either of his two grooved penises as she obligingly moves her powerful tail to one side.

in reproductive condition are dangerous beasts to observe and have even overturned small boats intruding on their muddy rivers.

Both sexes of alligators and crocodiles tend to bellow loudly in the spring breeding season. It's probably part of the mating ritual, but we can't be sure. They may also communicate by secretions from musk glands near their cloacas since females can lay a scent trail which males are easily able to follow. But, in fact, no one really knows how the sexes attract each other.

Once a couple gets together, however, the male initiates most of their courtship. He doggedly follows the female on land or in the water and may viciously attack any intruding competitor. With alligators, courtship may last several days as the male trails after the female, repeatedly clawing at her rump. Later he begins to rub his head against her throat, and in the water he moves under her and sends torrents of bubbles against her cheeks. Perhaps the most peculiar signal of all, however, is the strange little jets of water which he repeatedly sends in fifteen-second streams from his knobby back. It's a good trick, and no one knows how he does it. Finally, after all this, he arouses her primal instincts, and she begins her own mating behavior, crawling over and under him and bumping at his head with her snout. Both animals may now snort and grunt loudly and beat the water with their tails. With crocodiles, the females may display their excitement by rearing high out of the water before the male, jaws open wide and pointed to the sky.

Finally, when both animals are ready, the male, his behavior surprisingly gentle, clasps the female around the neck with his forelimbs and curls his body partly under hers. She obligingly lifts her tail and holds it away from him, exposing her cloaca. She may even turn slightly on her side to accept his penis which is now protruding stiffly from its cloacal pouch. The union is facilitated by the fact that his penis extends not only forward, but also to one side. It always erects in the same direction, so he probably always mounts from the same side. The pair lies joined for up to fifteen minutes. The union is a gentle one with very little thrusting or shoving, as the sperm quietly runs along its groove into the cloaca of the female. The male may be surprisingly faithful, showing attention to no other female for the entire breeding season. But he is hardly an ideal family man since, if he finds the eggs his mate will lay some two months later, he may lustily devour them all.

HORSES

A Snapping Vulva?

We all remember those old movies about the beautiful wild stallion who runs off with the rancher's best mares. In real life, the stallion may not be so beautiful, but he will run off with the mares. A stallion that has managed to acquire a harem, by whatever means, guards the mares, offers them some protection, and keeps them herded together. Since the stallion, then, is in constant contact with the mares, when they come into heat no introductions are necessary, and mating proceeds quite easily.

(As an aside, this is not the situation with donkeys. In this group, the "jacks" chase the "jennies" and fight with them until the females are subdued and willing to stand still when the jacks mount them. In fact, a jenny normally cannot copulate unless she is first violently subdued. Thus mules are usually the offspring of jacks and mares. The donkey stallion acts as if he has just stumbled across an unusually compliant female, whereas a horse, unaccustomed to fighting his mares, would have little luck with a jenny.)

Mares come into heat about every twenty-one days. At this time, some become edgy, uncooperative, and even dangerous, but others show almost no change in their behavior. The physiological changes in all mares, however, are quite marked. For example, as a mare comes into heat, her vulva swells and exudes a heavy gelatinous substance. In addition, she may behave rather strangely, standing straight-legged and urinating in quick spurts while spasmodically opening her vulva and snapping it closed in a most peculiar manner—a sure way to attract attention, one would think.

When she urinates, she may make a point of stepping in it, thus picking up her own hormone-laden smell. A stallion takes an inordinate interest in the feet of receptive mares and sniffs and snorts attentively around their ankles. After a good whiff, he may wrinkle his nose and expose his long teeth (in the manner of many plant eaters) as if he considered her scent disgusting (but he probably doesn't).

Mares have rather peculiar physiological systems. It's almost as if they have not yet been quite refined by evolution. For example, a mare may come into heat without actually having released an egg from her ovary, and conversely, she may release an egg without coming into heat. So, reproductively, the stallion could be wasting his time. But on the other hand, he probably couldn't care less.

When she is receptive, he approaches her and places his head on her back. If she doesn't indignantly trot away, he mounts her, holding her tightly with his forelegs. His penis erects rather slowly, since it is stiffened, not by muscles, but by the hydrostatic pressure of blood filling the many tiny interconnecting spaces of the organ. Basically, it is very similar to that mechanism which operates in men. The stallion, however, can boast of considerably more control over his penis than can a man. In fact, the horse can pull back the prepuce (the visible pouch) to expose just the head of his penis, and he can extrude or withdraw the shaft without ever attaining a real erection. You've probably seen these nonchalant "five-legged horses" standing around.

But once the stallion has mounted, he may thrust several times before he actually achieves intromission. When the penis enters the vagina, he continues to thrust, rupturing the hymen if she is a virgin and vigorously massaging the large clitoris nestled in a soft fold at the entrance. The head of his penis quickly swells, and the shaft almost doubles in diameter. He may ejaculate several times, but he stops rather soon and dismounts. In spite of their apparently incredible obsession with mares, stallions are rather quickly satiated compared to other domestic animals.

TICKS

Nose Job

Ticks, of course, have long been known as the scourge of American woodsmen, burying their hardened heads deep into the flesh and provoking endless conversations about how to get them out. And as a matter of fact, they can be quite dangerous. If a female implants herself along the hairline at the base of the human skull, the venomous saliva which leaks from her sucking mouth parts can paralyze the breathing center of a man's brain and bring about his death. To make matters worse, the rascals are hard to kill, and some of the adults can live up to two years without food.

The female, however, needs at least one blood meal in order to form her eggs. So she crawls, as a small mitelike animal, onto the skin of a bird, reptile, or mammal and pierces the skin of her host with her long snout, replete with backwardly directed hooks. As she engorges herself, she is oblivious to the small male creeping toward her. Completely unnoticed, he slips under her belly. He, of course, has sex on his mind, but he has a problem—he has no penis. And furthermore, she presents all the problems attendant to virginhood.

He has the solution, however, molded into his primitive nervous system by eons of evolution. Once he is in position under her, he wriggles his "nose" into her vagina! He then pushes and heaves and snouts about until finally he enlarges the orifice sufficiently. She, in the meantime, placidly continues sucking blood—unaware that she is about to be deflowered. It is no mean trick to impregnate a female when one has no penis, however. So what the male does after he has enlarged her vagina is to turn around so that his rear is under her and deposit a packet of sperm beneath her orifice. Then he turns back around, picks it up, and stuffs it as deeply as possible into her vagina by pushing it with his snout and tapping it with both forelegs. Finally, after much heaving, adjusting, fitting, pushing, and shoving, he is satisfied. It is in. He goes on his way. Later, grotesquely swollen with blood, she will withdraw her head, fall to the ground, lay her eggs there, and die.

MITES

Incestuous Obstetricians

Moth mites are tiny animals, related to spiders and ticks, which parasitize certain kinds of caterpillars. The female moth mite does not lay eggs but bears young which are virtually miniatures of the parents. It turns out that not only are the young born with the physical appearance of the adult, but they also appear with their sex drives in full swing. This combination makes for one of the most bizarre cases of sex among animals.

When a pregnant mite finds a caterpillar to act as host, she draws her sustenance from it and then gets down to the business of giving birth. She usually has several offspring of both sexes. When she gives birth to males, they linger around her vagina for a time, but they don't then leave, as one might expect. Instead, they bore into their mother's flesh to feed on her juices! Actually, they are able to survive alone quite nicely, but they don't leave her because they are waiting— waiting for their sisters to be born.

Soon the mother will again give birth. As soon as labor begins, the newborn males emerge from her flesh and congregate around her vagina which they watch closely. If the head of an infant female should appear at the opening of the birth passage, her older brothers scramble to assist with the delivery. One of them will push the others aside, turn around, stand on his head, grasp his sister with his nipperlike hind legs, and tug and pry until he pulls her out of the passage. (If a male should appear at the opening, he is ignored and will be expelled through the normal birth process. If a female has no older brothers waiting, she will be born the same way.)

But when a newborn female is finally tugged free, her older brother does not release her. Instead he turns, mounts her from behind, inserts his barbed penis into her tiny vagina, and inseminates her on the spot. The whole process of birth and fertilization is very brief. The little female, not even two minutes old and already pregnant, crawls away, perhaps a bit dazed by it all. But now she must find a caterpillar. Life is harsh for these little ladies. She has only a day or so to find a host before she dies of starvation. In the meantime, her brothers are home with the mother, sucking her juices and waiting for new baby sisters.

GORILLAS

You Put Your Foot Here and I'll ...

The huge male gorilla, resplendent with his silvered back and exuding the very essence of power and virility, has a penis about two inches long. Also, he is among the least sexy of all apes, being surprisingly difficult for a randy female to arouse. Because of his attitude, his sex life might not be very interesting if it weren't for his imaginative technique once he gets going.

First, we should draw a line between zoo animals and their wild brothers. Their sexual behavior is apparently quite different. For example, caged males masturbate, and, apparently, wild ones don't. And then caged animals have developed new patterns because the sexes are often separated by bars. For example, a male able to reach a female may respond to her proffered rear by poking a finger in and out of her vagina (but he may also do this in the wild). In the zoo, a female may choose unlikely sex objects. She may try to interest her keeper by placing his hand on her genitals, or she may spread her legs and attempt to pull him on top of her, much to his chagrin, especially if the Sunday crowd is watching. And then some captive females may mount baby gorillas and even terrified dogs.

If the sexes are allowed to mix, they may engage in extensive foreplay, mutually stimulating the gorilla erogenous zones. A female may fondle and lick the testicles of her partner, and the male may fondle her breasts and genitals. Then they sometimes wrestle, hugging, biting gently, and rolling around the floor. But when they actually begin to copulate, they usually begin by facing one another. The larger male often places one hand on the female's hip and swings her around, grips her hips with both hands, and enters her from the rear. Usually she draws her knees up under her and folds her arms beneath her chest, which is pressed against the floor. The male, squatting behind her, may continue to hold her hips, or he may dangle his arms nonchalantly from his sides. In some cases, before he is able to enter her from the rear, she may pull away and fall onto her back, legs apart, and accept him this way. In this case, the male squats between her legs or stands over her on all fours. In rare instances, a male facing the female may simply lift her entirely off the ground and do it standing up as she grips his hips with her legs.

Wild gorillas have rarely been caught in the act. After all, they are shy creatures, inhabiting some of the remotest parts of Africa, and few observers are disciplined or curious enough to travel with them through their steaming jungle home. The reports which have come in, however, indicate that the sexual behavior of wild gorillas is much more direct than that of captive animals.

Wild gorillas live in groups, browsing placidly through their forests, sleeping a lot or just sitting around. A group is dominated by an old silverback male, and his authority is rarely challenged. However, females may solicit the attentions of any male, and their amorous adventures arouse only slight interest from the old male, or any other group member, for that matter. One pair was observed copulating on a slope, emerging from the bushes with the female crawling along on her knees and elbows, and the male holding her hips and shoving from behind as they made sporadic progress down the hill. The female cleared branches out of the way as they went. After they traveled along for about ten minutes, their soft oooo oooo's became harsher, and the female began emitting short, piercing screams. Their downhill progress finally stopped as they came to rest against a tree trunk. The male then gave a hoarse, trembling sound from his protruded lips, a gasp which gradually became a roar. Then he withdrew from the female and sat back as she turned and walked slowly back up the hill.

Wild gorillas do it from the side, front or rear, with the male on top, the female on top, both lying down, or the male sitting with the female on his lap. In this position, he may thrust against her or hold still and rock her back and forth. In some cases, the female determines the position. For example, if she finds a male lying on his back, she may fondle his genitals until he achieves an erection and then squat over him, inserting his penis into her as she thrusts with her hips. Her clitoris is much like that of women and undoubtedly provides sensory reinforcement.

A male gorilla may be imaginative, but he is not particularly adroit. He may have to thrust about a hundred times before the small black penis, stiffened with a slender bone, reaches its goal. It is easy to tell when he finally succeeds because they both begin to pant. A scream, roar, or sigh signifies the climax of the session.

TARANTULAS

A Parting Pinch

In wolf spiders, the group which contains the tarantulas, the female is larger than the male—and hungry. So the male has the predictable dilemma—how to avoid being devoured by his mate.

A female wolf spider may live twenty-five years—as long as a seventy-ton blue whale. The male doesn't live as long, even if he manages to avoid her gaping jaws, but at least he doesn't have to risk his life in his first years. In fact, he shows no interest in breeding until he is about eleven years old. But he may not get much older unless his instincts have matured properly.

When he is ready to mate, his first problem is how even to approach the female. The problem is solved through visual signaling. Sometimes the female signals and approaches the male, but often the task falls on him—and so does she. His signaling is usually accomplished by waving special arms, semaphore fashion. After long pauses and cautious steps, punctuated by mesmerizing arm waving, he finally approaches her. In some cases, she doesn't even seem to be aware of his presence. In some species, if she fails to respond to him, he may have to slap her around a bit to get her attention. In other species, she is *quite* aware of him, and when he is close enough, both may rear up and engage in "forearm play," touching each other lightly with their hairy forelegs. After a bit, the male seems to come to his senses, and his

Among tarantulas the male may be killed as he approaches the female to mate. If he succeeds in mounting her, since he lacks a penis, he must rake sperm from his arm into her vagina, as the male at left is doing.

hair literally stands on end. He is in no position to dally; he leaps onto her back.

Alas! The male spider has no penis. So he, like the octopus, uses his arm. He had earlier spun a small web and deposited his sperm on it. Then he spent the next couple of hours tediously raking the sperm into special pouches at the tips of two of his forelegs. Since these would be his penises, and since he wouldn't have much time, he lubricated them by drawing them through his mouthparts. Now, astride his deadly bride but with their bodies reversed so that his head is at her rear, he must reach under her and somehow slip his sperm into her vagina. She now lies flattened against the ground, her belly compressed against the surface. His chore is made more difficult because she has essentially two vaginas, side by side, each covered by a hard plate. So at this point, she becomes more helpful; as he leans over to one side toward the nearest opening, she twists so that it is easier for him to reach. He then rakes his arm several times across the hard, jutting covering of that opening until it "catches." The lid is raised, and he carefully packs the vagina with the sperm from the tip of his arm. Throughout the act, his body rises and falls rhythmically, perhaps a way of calming her. When he is finished, he moves to the other side. She twists again, and the process is repeated. Each opening may be serviced several times as the sperm disappear into her bulbous body. Finally, though, she becomes restless and starts to move about. It is time for him to be going along.

Some males just leap off her back and run for it. But they are often caught and overpowered, their bodies to be macerated and sucked dry by their lovers. Other males, however, take no chances. They reach over with one claw and suddenly clamp down tightly on the female's rear. Then they flatten out and with a mighty spring, leap through the air, carrying her with them a few inches and bowling her over. While she rolls up in the typical defensive posture of jumping spiders, the male, who has landed some distance away, clears out. Those males which give their ladies a little pinch on the way out live to breed another day.

Other jumping spiders solve the whole problem entirely differently. In some species, the male confronts the female directly and grasps her jaws in his, holding them immobile while he reaches under her with his sperm-bearing arm. In another species, the male leaps astride the female and quickly ties her down by spinning silken threads back and forth over her body. Then he takes his time, deposits his sperm, and by the time she has struggled free, he is long gone.

Once he begins breeding, his life is short in any case. A male will only be able to repeat his sexual adventure three or four times at the most. Then he begins to weaken and actually seems to shrink. He is no longer bold. His time is finished. He crawls feebly from place to place, his hairy arms trembling, until he finally stops for the last time and dies.

LOBSTERS

Roll Me Over

In order to understand how lobsters do it, we have to have some idea of how they are built. We know that a lobster has no neck, the head area being fused to the back forming a stiff covering shell. Underneath this shell, we find the huge pincers (in some species) and four other pairs of long walking legs. The flexible abdomen gives rise to five pairs of tiny paddlelike legs called swimmerets, and the flattened "tail" brings up the rear. Notice that we didn't say anything about a penis or vagina—because there are none. But perhaps the strangest thing is that although the female is raped, lobsters *really* don't do it at all!

What happens is, in the fall breeding season when the male finds a female, he plays across her body with his long antennae. If she doesn't retreat at his light touch, his demeanor changes. He grasps her rather clumsily with his huge claws. A disjointed and uncoordinated battle then ensues as she seeks to escape while he tries to roll her over onto her back. The awkward beasts, encumbered by their heavy protective outer skeletons, surge back and forth across the ocean floor until finally the stronger male prevails. Once she is on her back, he pins her securely by grasping her large claws in his and holding as many of her other appendages as he can reach with his walking legs. Then he cups her tail with his large abdomen to steady her and slips his first pair of walking legs into two pockets between her walking legs. This is the lobster's version of copulation. His walking legs are grooved by sperm channels leading from the large testes under his shell. Once his legs are inserted into her special pockets, he may hold her immobile for as long as an hour, his sperm continuously running from his testes and down his grooved leg until her pockets are filled. Finally he withdraws his legs from her body and releases her. She quickly rights herself as he walks stiffly away.

She will store the sperm through the winter, and only in the following spring will the eggs be fertilized. She prepares herself for the event by turning onto her back and cleaning herself, removing all debris from her abdomen. While on her back, she rolls from one side to the other on her rounded shell and exudes a sticky substance from special glands under her abdomen—a thick fluid which covers her swimmerets. While she is still upside down, the eggs begin to appear at an opening near her third walking legs. As they move down her abdomen toward her sticky swimmerets, she releases the stored sperm over them. The fertilized eggs then cling to these small, sticky legs. When the eggs have been released, she rights herself and walks away with her new burden. She will carry the developing embryos around for the next several months, gently fanning oxygen-laden water past them as they change slowly in the cold, quiet waters.

When the young hatch, they emerge from their egg cases headfirst and would helplessly tumble away from the mother were they not anchored by a thread running from their tails to the inside of the old egg cases. Even though they immediately reach up and pinch the threads in two, they will cling tenaciously to their lifelines for the next several weeks, only gradually building the strength and courage to leave their crusty old mother.

CATTLE

And Mr. Coolidge

Cattle probably have the worst sex lives of almost any animal. There are a lot of pregnant females around, it is true, but most of them have very likely been impregnated with the chilly instruments of man. Artificial insemination is the rule in these economically important animals. Bulls are tricked into copulating with dummies (not dim females, but heavy wooden frames) which they come to accept as sex symbols. The bull's semen is collected in a receptacle and later placed into the uteri of selected females.

Sometimes, however, he won't mount the frame unless he has first been excited by a live female in heat. Interestingly, if the same female is used to arouse him several times, she loses her magic for him and he may refuse to mount the frame, so a fresh female has to be brought in to revive his waning interest. In other cases, a sexually exhausted bull in the presence of a familiar cow will remount a frame which itself has been altered in some small way so that he takes it to be a new female. (This rejuvenating quality of the "new female" is familiar to many animals from insects to sex therapists. It is called the "Coolidge effect" since, as one story has it, President and Mrs. Coolidge were being given separate tours around a farm. The First Lady commented that the rooster in a yardful of chickens must keep quite busy and suggested that the farmhand tell that to Mr. Coolidge when he came by. The farmhand dutifully told the President that Mrs. Coolidge had asked him to point out that the rooster copulated many times each day. "Always with the same hen?" Mr. Coolidge asked. "No, sir," came the reply. "Well, tell Mrs. Coolidge *that!*" said the President.)

Under natural conditions, bulls will fight over females by engaging in head-to-head pushing contests, one suddenly wheeling and loping away, leaving the female to the stronger or more determined animal. The cow herself is a randy animal when she comes into heat (about every twenty-one days), however, and the bull has no trouble interesting her in a bit of a frolic. When a cow is ready to mate, there is no mistake about it. She will mount other cows, or even the bull himself, to make her point. But the bull may not be able to mount her until he has gained her "permission" by first placing his head on her back. If she doesn't move away, permission is granted—she is willing.

Actually, her willingness is important because periods of heat are unusually short, only lasting about fifteen hours. They can't waste time. Cows are also unusual in that their eggs are released *after* their heat period, rather than during it.

Cows almost always have thick mucus plugs in their vaginas, which tend to impede the penis. But at their breeding times, the plug partly dissolves, and the vagina, a long, rough, and rigid tube, becomes wet and slippery. The vulva is loose and fleshy and houses a well-developed clitoris, the end of which peeks into the vaginal tube.

The penis of the bull is not like that of a horse or man. It is largely nonerectile, and its tough, fibrous makeup does not permit it to swell much. Instead, the unextended penis is curved into an S-shape, held flexed in the body by two powerful retractor muscles. When these muscles relax, the gleaming red organ immediately springs forth in all its splendor. At least the cows seem to appreciate it.

When the cow first approaches the bull, he immediately begins to nose at her rear and lick her vulva. His penis may then become partially unsheathed and begin to drip semen.

From time to time he may throw his head back and curl his lips up in the sexual grimace of many hoofed animals.

When he mounts her, his penis quickly extends to its full length. The bull's penis is long and narrow with a pointed and slightly twisted tip. Because of its size and shape, it readily slides deep into the vagina, and he quickly ejaculates a small amount of high-density semen far up near the uterus.

The cow has remarkable control over her vaginal muscles, and she squeezes his penis at two places, near the uterus and far back at the mouth of the vagina. The vagina contracts strongly the instant before the penis enters it, producing a tight pressure on the penis which undoubtedly helps to excite the bull. In any case, something excites him because he is able to ejaculate up to eighteen times in a single coupling.

In other animals, such as the dog, horse, or human, the stages of ejaculation are sequential, so it is easy to describe what happens in some order. But the bull gives only one pelvic thrust, and then everything happens at once; thus the stages cannot be easily defined. What happens is, the sperm-laden tubes leading from the testicles contract, forcing the sperm outward where it is mixed with secretions from various glands, such as the prostate. At the same time, the tubes leading to the urinary bladder close tightly so that the semen does not take the wrong route, and the muscles of the pelvic region contract sharply, forcing the sperm out the tip of the crooked penis. Apparently the glandular secretions do not activate the sperm, as was once believed, but simply wash the sperm from the tubes.

About 280 days later, the cow will give birth. She will nourish her young with copious amounts of a modified sweat. Some of the excess will be taken by humans, heated, irradiated, and sold in paper containers.

MOSQUITOES

Whine Not

As the male mosquito changes from the "wriggler," which has been writhing about on the surface of some pond, into an adult insect, his rear end turns around—the last two segments of his abdomen turn a full 180 degrees. This twisting is necessary for him (and most flying insects) to be able to copulate. Now wearing his penis upside down, he is ready—he goes in search of a mate.

The male mosquito doesn't rely much on charm and persuasion. When he finds a female, he immediately leaps onto her back and clutches at her with the little grappling hooks on his feet. Then, by a series of intricate steps, he swings under her so that they are now face to face, his body suspended beneath hers. However, her vagina is still shielded from his probes by two hairy, paddle-shaped plates. He has an arsenal of gimmicks to meet such problems, though, and two pinching claspers protruding from his rear reach up and grab the two hairy plates, causing the plate under her anus to jerk upward, exposing the edges of her vagina. James Bond has nothing on him. Now two special anal hooks reach up and pull the female's genitals downward toward him. At this point, he extends his bizarre penis, all abristle with jagged teeth. These teeth, dangerous as they seem, actually work as a kind of key; they mesh with an opening on the upper surface of her vagina, causing it to open like a valve. When it is wide open, a direct channel is exposed which runs from her vagina into her bursa, a sac where she stores sperm. In spite of the intricacies of his task, he wastes no time. As soon as his bristly key has opened the valve, he immediately ejaculates a surprisingly large amount of semen (holding about 2,000 sperm) into the bursa.

He then disengages himself and lets her fly away. She has one final comment about the whole episode, however. Just as the victorious male pulls away, she gives him a swift kick with her hind legs. He recovers in midair and flies away to spend his few remaining days harmlessly sucking the juices of plants. She will outlive him, and she immediately begins her search for that blood meal with which to nourish her developing eggs. Thus you can rest assured that the delicate lady sucking your blood is not a virgin.

SNAKES

A Left! And a Right!

If you should walk up on a pair of copulating rattlesnakes and they try to slither away, the stronger snake will drag its mate along by the anus (actually, the cloaca). They don't uncouple very easily. In fact, vipers and adders sometimes form huge amorous balls, the pairs joined anus to anus, with none able to extricate itself. Bold, if addled, foresters have been known to pick up the whole fornicating lot and carry them away.

The reason the snakes aren't able to separate is because of the horrendous design of the snake penis. Its end is soft and pointed, but its base is a forest of stiff, backwardly directed barbs.

It obviously takes a bit of persuasion before a female will allow a male to penetrate her with any such device. So the male often must go through the ritual of foreplay. The erogenous zones on a snake seem to be under its chin and around its cloaca. Once the male has found the female (which shouldn't be too hard to do since both sexes have powerful musk glands opening near the anus), he may begin things by rubbing his sensitive chin along her back, which probably turns him on more than it does her. She may later get her own chin stroked, and ultimately they will stimulate each other's anal openings. Different species, of course, have their variations. Boas and pythons, for example, have tiny rudimentary "legs" protruding from near their tails which males use to scrape at the anal areas of females. This makes a crude rasp-

Male snakes have two barbed penises that evert like the fingers of a glove. They can copulate with either one. If these Australian brown snakes were disturbed, the female might try to run away with the male hanging on in a most undignified fashion.

ing noise which can be heard several feet away, but it seems to work. The female responds by rolling a bit to one side so that the male can reach her cloaca. He may help her along by slipping his tail under hers and lifting her rear end.

At copulation, the male lies alongside or slightly under the female, and at the moment of truth he extrudes his bizarre penis from his cloaca. It is now evident for the first time that his penis is not one organ, but two! And both are fully potent and barbed. The one which is actually used simply depends on which side she is on. The penis itself is not a tube, by the way. Instead it is essentially a fleshy, grooved organ, and the sperm flows along the deep channel into the female. He probes her anal area with the nearest penis until he finds the opening, whereupon he quickly inserts it. The penis is erected by turning inside out, and because of the hooks and barbs which hold it in place, it must be withdrawn by carefully reversing the process. It would obviously be in poor taste just to jerk it out, even if he could. The snakes are in no hurry, though, and they may lie joined together, occasionally thrusting or undulating, for an entire day.

The females are able to store the sperm, somehow managing to keep them alive almost indefinitely. In fact, some females have given birth up to five years after their last affair.

EELWORMS

The Living Vagina

Eelworms are tiny parasitic worms which thrive on some plants and a few animals. They are actually rather unremarkable little creatures, except for their sex lives. Their sex lives are incredible!

To set the stage, we should point out that the sexes are quite distinct. The female is larger, and her vagina is in the center of her body. The male is equipped with a pouch at the end of his body which encloses a forked hook that is introduced into the female's vagina during copulation.

Actually, there are a number of kinds of eelworms. In the turnip eelworm, sexually mature females look like tiny lemons. They imbed just under the skin of a turnip, and when they are ready to copulate, they must break through the rind. They don't crawl out in search of a mate, though. They just stick their vaginas out. The males, which have been comfortably living in their own little caverns in the same turnip, emerge and crawl around the surface of the rind, poking their little hooks into any vagina which appeals to them, holding it fast, and filling it with sperm. The male's task is finished, but the female must remain in her cavern until her young can properly develop. Then her cavern becomes her tomb.

And then there is the strange case of the bumblebee eelworms. They copulate in the damp earth, whereupon the male promptly dies. The female, on the other hand, lives on and crawls through the earth on an unlikely odyssey—seeking out hibernating bumblebees. When she finds one, she penetrates its body wall and crawls deep into its tissue. Once she has nestled in, a strange transformation begins. Her *vagina* begins to grow. It looms larger and larger until it absorbs the ovaries and the uterus. Then it continues growing until it has dwarfed the entire body. It is now a gigantic organ about twenty thousand times as large as the worm itself! Early investigators found the huge cystlike vaginas often dangling a small thread which they assumed to be the male. But they were wrong. It was the female's body. Finally the great vagina casts off the tiny, useless appendage which gave rise to it. The monstrous organ then leads a life of its own, taking its sustenance from its host until, finally, it gives birth to a new generation of tiny young. The little worms crawl out of the bumblebee's body in the autumn, leaving the withered organ which was their mother to die in her winged tomb.

WASPS

Taking Her Out to Eat

Wasps boast a wide variety of social systems. For example, there are "solitary" and "social" species. The social wasps share large, communal nests, their lives quite unlike the lonely territorial vigils of the solitary species.

Among solitary wasps, the males hatch first and stake out territories which they vigilantly patrol, pouncing on almost anything that intrudes on their air space as they wait for the females to hatch. In other species, a male may be unconcerned with a territory per se but lays claim to some female who already has her own nest, like a would-be gigolo. When he finds such a female, he enters her nest and darkly lurks about inside, irascibly defending it from any intruder. Whenever the female enters the nest on her daily rounds, he immediately leaps on her back and copulates with her, his entire attitude clearly demonstrating the lack of poetry in his soul. This may go on for days or until she signals her unwillingness by twisting away from him. She probably has no idea that he will take her refusal so hard; it kills him. Actually, she is only reflecting the end of her mating period, but this also means he has no further usefulness, and his body begins to weaken until finally he dies.

During their active sexual lives, though, active is the word for it. The male usually mounts the female from above, straddling her body and holding her with his jaws and legs. He may approach her on the ground, but he often nabs her in flight, forcing her to land and then copulating with her for perhaps an hour. He is not entirely without grace, however. While he is astride her, he may stroke her antennae lightly with his mouth parts.

In order to copulate, he bends the tip of his abdomen sharply downward and moves his stinger out of the way (or retracts it). He erects his penis by forcing fluids into it from his abdominal cavity and then inserts it into the female's vagina, locking it firmly in place. After that, things proceed differently in different wasps.

In the digger wasp, once the pair is firmly joined, they may take to the air, nonchalantly flying from flower to flower, the male continuing to copulate with her in midair.

In the wasp called the cicada killer, once the male has locked in his penis, he twists off the female's back and faces the opposite direction. If they are disturbed, they both take to the air, but since the female is larger and stronger, she ignominiously tows the male around backwards by his penis until she darn well decides to settle again.

In yet other species, the female is wingless, so the male tows her around (upside down and backwards) for an hour or so, their genitalia firmly interlocked. However, the male may be so kind as to transport her to some place where she can grab a bite to eat while being fertilized. He may even feed her himself by regurgitating food for her. Any kindnesses are deeply appreciated because these females are the dependent sort—short-legged and practically blind. In fact, since the females are so helpless, the lift provided by the winged male may be the only means of scattering the species.

DRAGONFLIES

It Was As Long As Your Arm

The dragonflies of today are mere miniatures of the great aerial beasts which clattered through prehistoric swamps. They were truly monsters in those days, their huge bodies almost a yard long, and they threatened a whole host of other animals. Fortunately, today they are smaller, and they terrorize mainly mosquitoes. Actually, two kinds of dragonflies exist today, the damselflies and the "true" dragonflies (in the latter, the two pairs of wings are of different sizes, and they fold parallel to the body instead of perpendicular to it).

The reproductive structure of the male dragonfly is unique in the insect world. In most insects, the male's penis simply ejaculates directly into the female vagina. For some strange reason, and through unknown evolutionary processes, dragonflies have complicated matters. The male's genital opening is at the usual place, near the tip of his long abdomen; however, before he mates he exudes his sperm from here, and then, doubling his long body under, he deposits it in a complicated organ far up toward his midsection. The sperm is stored here until he locates a mate.

When he spots a female flying along in his territory, he quickly runs her down and lands squarely on her back. Once she recovers, they fly along together as one. Then, holding her with his legs, he tucks his abdomen under himself, and with the two hooked claspers at its tip, he grasps the back of her

The male damselfly first deposits his sperm into his own body, near the midsection. With the end of his tail he then grasps a female behind her neck. Once he has her, she swings like a trapeze artist, coupling her vagina to his sperm receptacle. This is when fertilization actually takes place.

neck or the top of her head (depending on the species). Then he lets go of her with his legs so that she hangs by her head from the tip of his rear end.

Like a trapeze artist, the female swings her own abdomen up under his so that her vagina, appropriately located at the tip of her abdomen, contacts the sperm-storage organ under his body. Then, through a complicated process, the sperm are pumped into the vagina.

The female may then break away from the male to lay her eggs (but the male, for some reason, usually follows her), or the pair may continue to fly joined and may land in tandem on a reed. When this happens, the female may walk right down into the water (breathing from air bubbles trapped along her body), dragging the male along. In yet other species, the male may fly along the surface of the water, still holding the female by the back of the head, repeatedly dipping her tail into the water and washing off the fertilized eggs so that they can begin their aquatic period of development—a period which may be as short as six weeks or as long as six years! The nymphs which hatch are terrible predators, attacking everything in sight, including small fish as well as each other. To lessen their general appeal even more, true dragonfly nymphs swim in a rather indelicate jet-propelled manner by squirting water out of their anuses. So if another animal in the pond is not devoured by these creatures, it is surely grossed out.

We don't know how such a peculiar mating system came to be, but since only predatory animals, such as octopuses and spiders, copulate by such indirect methods, perhaps the technique arose as part of the male's self-defense system. Maybe the female is so busy figuring out the next step, she forgets what a tasty morsel he might make, but more likely the gymnastics just result in her never being in a position to behave as a predator. One just can't be too intimidating while hanging from a penis.

CHIMPANZEES

Families That Play Together . . .

Humans and chimpanzees are alike in many ways (for example, our blood hemoglobin is indistinguishable), but we are very different in our sexual behavior. In spite of our differences, next to us chimpanzees are clearly the sexiest of all the great apes. In fact, chimpanzees clearly show how wrong the notion is that copulation in nature is "reserved" for reproduction.

Chimpanzees, first of all, are highly promiscuous. A receptive female may be mounted by every male in the group in rapid succession. Just how rapidly depends on how appealing they are. Males consider some females much more desirable than others, and the "sexiest" individual may be one which to human eyes is also the ugliest. This promiscuity means that even the very young get in on the sex. Old females may squat low to the ground so that infant males can reach their vaginas with their tiny penises. Whereas infant females rarely copulate, adolescent females are often mounted by adult males.

Female chimpanzees menstruate about every forty days, always at the end of their most receptive periods. When a female is receptive, the skin around her vagina and anus swells grotesquely and turns bright red, a sure signal that she is ready to mate. (She is then called a "pink lady.") It should be made clear, however, that her sexual activity is not *restricted* to these times. She may readily copulate when there is no chance at all that she may become pregnant. The chimpanzee's extreme promiscuity probably functions in maintaining a high level of interaction within the group, and a pleasant sort of interaction at that. Perhaps the bonds between various individuals are also strengthened in this way—an important factor in a highly social group. The idea, then, is that sex keeps them together and promotes goodwill.

A female may solicit the attentions of a male by approaching him, turning around, and crouching low to the ground with her rear end raised and her vagina easily accessible. This is the same position that she assumes when he signals to her that he wishes to copulate—as he very often does. But if he has not requested her favor, he may not respond to her enticement. He may instead closely inspect her vagina, sniffing at it, or slip his finger into it and then sniff his finger.

A male may request a female's favor in a number of ways. He may swing noisily through the trees, dropping to the ground and rattling and shaking small saplings. He may also stand with an erection, looking directly at a female and rocking back and forth on his feet, or he may sit there and simply raise one arm to beckon her. Usually the hair on his head, shoulders, and arms stands on end and emphasizes his message. If a nearby female doesn't immediately respond, he may reach out and touch her back lightly, whereupon she quickly crouches and accepts him.

His slender penis is a few inches long and is stiffened by a bone. Some males have tiny knobs covering their penises (which may aid in the stimulation of the female). In any case, once his penis is inserted, he makes about a dozen powerful thrusts with his hips and ejaculates. While they are copulating, the pair may remain silent, or he may grunt softly while she squeaks or even screams. When his penis is withdrawn, it jerks up and down spasmodically until he loses his erection.

For some reason, chimpanzees tend to interfere with each other's sexual behavior. Adolescent males are particularly nervous about copulating in the presence of higher-ranking males and only sneak in a quickie while the dominant males are gone or otherwise preoccupied, such as in napping or being groomed. A copulating pair particularly upsets infants and adolescents. They often scream and try to push the male away, or they may jump on the female's back and scream directly into his face. The entire sex act is somehow particularly upsetting to the offspring of a copulating female. And the young chimpanzees clearly regard the male as the culprit.

KANGAROOS

And Slimy Little Slugs

When the first kangaroo came to London at the end of the eighteenth century, people paid the rather hefty sum of a shilling apiece to see it. They might have paid more if they had known how it reproduces.

Whereas kangaroos may be rare in London, in Australia they are slaughtered by the thousands for practically any reason, from target practice to dog food. In spite of their abundance, however, scientists were for years baffled about how the baby got into the pouch. Did the vagina open into some kind of false bottom in the pouch? No one could find it. In fact, the problem was not solved until this century. A zoo keeper just happened to be looking at a kangaroo's vagina one day when, to his surprise, he saw a slimy little slug crawl out. The female was sitting on her haunches with her tail under her and her great hind legs sticking straight out, as a man might sit on the grass. As the keeper watched, the little slug crawled and squirmed its way upward and finally disappeared into the pouch. The baby was in, and the secret was out! People began to keep a close eye on kangaroo vaginas.

Subsequently it was found that some females pave the way for the tiny embryo by licking their fur, forming a moist pathway from the vagina to the pouch (although she may only be cleaning away birth fluids). Others, however, pick up the tiny infant in their lips and deposit it in the pouch. In all cases, the female kangaroo sits down and protrudes her cloaca out and upward so that the young have a shorter distance to travel.

The male kangaroo's penis lies behind his testicles, so it points downward when "erect." The female has two vaginas, but a third, central one will appear when she gives birth. The birth process startled the biological world when it was discovered by an idle zoo keeper.

As soon as an embryo reaches the pouch (which takes less than thirty minutes), it attaches to one of the four nipples it finds there. The nipple immediately begins to swell within its mouth, and the embryo is held fast for the next several weeks. The attachment may be important for a number of reasons. The infant could be jostled out as the mother bounds across the plain. And then the males, for some reason, tend to pluck the young out of the pouch and toss them away to die. However, all in all, the pouch is a secure place. Even the plumbing is specially adapted for the infants. For example, the mother has special muscles with which she may actually pump her milk into her ill-formed and weak offspring.

The female kangaroo is a sexy beast and will come into heat even while she has a baby in the pouch. When she wants to copulate, she begins to follow some male around, staying right after him and nuzzling him whenever he stops. When he turns to her, of course, she looks startled and bounds away. But he is not to be fooled by her sudden coyness and bounces determinedly after her. If she is already carrying an infant, she will draw the mouth of her pouch tight by a special set of muscles during the chase. Finally, she gives in and stops running. As he approaches, she crouches forward and moves her heavy tail to one side, exposing the shallow depression under a moon-shaped flap, which is the opening to her cloaca. The male clutches her around the waist with his small forepaws or places them lightly on her shoulders.

His penis lies *behind* his scrotum and projects downward and backward. So when he enters her from the rear, he must thrust downward. When he ejaculates, the sperm divides into two groups because her tract divides into two vaginas, one on either side. When the infant is born, however, it will travel down a central vagina (closed during copulation). Even the sperm of marsupials is peculiar in that many of them are joined in pairs, fused at their heads. As soon as the male ejaculates, he withdraws, and the two animals rejoin the grazing herd.

But the rest of the story is also a bit strange. The new baby will develop normally until it is composed of about one hundred cells—about the size of a pinhead. Then it goes into a state of suspended animation. It will grow no more until the older baby in the pouch either dies or grows up and leaves home. Fortunately for it, most babies die in their first year, mainly because of drought and human-caused stress. If the baby survives, it may be around for a long time since it must remain attached to the nipple for as long as eight months, and even when it is almost as large as the mother, it returns to the elastic pouch each night after a day of grazing with the herd. It's a ludicrous sight indeed to see a large head, protruding from a bulging pouch, snatching at grass as the mother hops along.

DEEP SEA FISH

Togetherness

Deep sea fish live in what is, from our point of view, a most peculiar place. And they have had to contrive numerous bizarre characteristics in order to survive there.

We really know very little about them, but it seems that they are often carnivorous and have surprisingly large heads and jaws for their size. The large jaws, of course, would help in a world so sparsely populated that a carnivore has to be prepared to eat whatever it comes across. The larger its jaws, the more kinds of things it can eat.

Not only is food hard to find down there, but so are mates. In fact, for years trawlers brought up only females of such grotesque species as fishing frogs and angler fish. It was noticed, however, that they were often parasitized by tiny sucking fish. It was not discovered until rather recently that those miserable parasites were the males of the species!

In hindsight, of course, it makes sense. It's hard to find a mate in the blackness of the deep sea, and when a pair finds each other, they'd best "stick together." It turns out that the large, sluggish females of such species constantly emit brief flashes of light from organs on their sides and great heads. As the pale blues, yellows, and oranges pulse in the blackness, tiny, fast-swimming males are attracted to them. The males are distinguished by their small size, their peculiar viselike jaws, and their outsized organs of smell (which may help them to locate the flashy ladies).

Once a male finds a female, he attacks her in a very literal

The male of this deep sea fish is reduced to the tongue-like append-age attached near the female's rear *(left)*. He is totally dependent upon her and is useful only as a sperm-producer.

sense (although she may be twenty thousand times as heavy as he is). He bites deeply into her flesh—the last bite he will ever take—and thus attaches himself firmly to her side, her head, her gill covers, or whatever. Once he has established himself, he truly becomes one with her as the cells of his jaws and tongue begin to change and the flesh of the two animals fuses together. He no longer has need for a circulatory system; her blood delivers his nutrients, and his digestive tract, useless now, withers and disappears. He may help get his own oxygen by sucking in water from tiny openings at the sides of his mouth, but he is, in sum, almost totally dependent on his bride.

As everything else withers, his sex organs begin to develop to the extent that he virtually becomes a living sperm producer, unable to survive alone. But her tolerance of his miserable presence means that when she is ready to lay eggs, she doesn't have to search for a male to fertilize them. When she lays the eggs, the hormones in her blood induce him to release his sperm. The price she pays for this service is not much. He is so small that he is certainly not a large drain on her reserves, and not much of what she catches needs to go toward his sustenance. His role in the family has been minimized to the ultimate. Only those fish species which are comprised solely of females could have subjugated the male role more.

FROGS

Not Really

Frogs don't really do it. Actually, though, that's not quite right because one species sort of does it. When the male tailed frog finds a female on the bottom of a pool, he grasps her firmly just in front of her hind legs and presses his cloaca near hers. He then protrudes his cloaca as a kind of penis and thrusts it into her opening, squeezing sperm over the eggs within her body. She will later lay the eggs in a lumpy, gelatinous string, or "rosary."

Most frogs and toads, however, practice external fertilization. In other words, the female lays the eggs while the male, clasped tightly to her back, pours sperm over them. But there are a number of variations on this theme, and some of them are quite unexpected, if not bizarre. Frog sex is not as simple as it might seem. For example, it appears that female frogs experience something like an orgasm, but perhaps a general kind of climax, one involving the whole body.

Some frogs might have problems with their sex because they don't know what it is. In tadpoles, the developing reproductive glands have an inner "core" and an outer "rind." If the core develops at the expense of the rind, the frog will become a male. The opposite trend produces a female. Half the tadpoles usually swing one way, half the other. But if the weather is cold, most will become females, for some reason. Once frogs reach adulthood, however, their sex is quite well established (although you might not think so to watch males in the breeding season attempting to grab everything in sight,

Frogs mate in a variety of ways. This male tree frog is clasped firmly to the female's back, held fast by special fingers. As eggs emerge from her body, he releases sperm over them. Other frogs, however, may practice a rudimentary form of copulation, complete with orgasm.

including each other). So let's consider some examples from the frog world.

A male midwife toad calls hoarsely from a little hole in the ground and attracts any females who happen to be within hearing distance. However, once they locate the burrowed rascal, he won't come out. It takes a lot of enticing, but once he emerges, he loses his shyness and pounces on the female, gripping her tightly around the neck. Then he begins gently to massage her cloacal opening with his long hind toes. This triggers her egg laying, and she begins to discharge great jelly-like masses of eggs in explosive spurts. The male is ready for them, and as he releases his sperm, he cups his webbed feet underneath her cloaca so as to catch the eggs as she expels them. Then he begins to kick his legs so that the egg mass becomes wound about his ankles. He carries his soft shackles for several weeks until he finds a pond not already occupied by other tadpoles. Here he gently scrapes off his burden so that his offspring can develop in their new home, free of competitors.

Not all frogs, by the way, come from tadpoles. The female nest-building frog lays her eggs under stones or logs while clasped by her suitor. These eggs, however, develop directly into miniature versions of their parents. The tiny Andean Darwin's frog sees its first days as a fully formed frog also, but it goes through a most peculiar sequence before it begins living on its own. The female lays the eggs on the ground, and the male fertilizes them in the usual way, clasped to her back. The female then returns to the water, but the male lingers with a group of his colleagues who have been attracted to the scene. They will watch the eggs for several days. When the eggs are about ready to hatch into tadpoles, each male picks up a number of them and slides them with his tongue deep into his vocal pouch. Here they will remain in the safety of his throat until they crawl out of his mouth some days later as little frogs.

In the little poison frog, the male carries the tadpoles on

Breeding activity may become so intense that males will clasp almost anything from a man's boot to each other. Here, two males have found a receptive female and both are attempting to fertilize her emerging eggs.

his back and deposits them later in some quiet pool. But in order to acquire his charges, he must first, of course, acquire a mate. So after a morning rain, he sets out, hopping around and buzzing softly. As he moves about, he is sure to attract an entourage of ready females. They follow after him closely, some actually jumping squarely on his back. He seems to ignore their suggestive behavior and he continues to hop about, but not so fast as to lose them. Finally he dives under the leaves, and a female immediately dives under the cool blanket after him. No one knows exactly how they do it because they do it under the leaves, but after they emerge, the female goes her way. He repeatedly visits the safely hidden eggs for the next several weeks, and when the tadpoles hatch, they crawl aboard his back, and he carries them to their new home in the pond.

In other species, the young actually develop while on their mother's back. As the male pours his sperm over the emerging eggs, he catches the jellylike egg mass and smears it over the female's back. The eggs become imbedded in little pits or folds (or sometimes in one large sack) in her back, and here, in relative safety, they complete their development. In one species, the Surinam toad, the female extends her cloaca as far as possible so as to assist the male in getting the eggs to her back, where they stick fast. Each egg seems to act as a strong irritant because it raises a large welt on her skin, sinking into its own mound of maternal flesh and sealing itself in except for a small slit through which a tiny toad will emerge some time later. The general advantage of parental care is indicated by the fact that such animals lay far fewer eggs than species which leave everything to chance. Presumably each cared-for egg is more likely to survive, so fewer have to be laid.

In some cases, the means of getting the eggs to water seem almost ingenious. In one tropical tree frog, the male clasps the female as she climbs around the trees on her long, sticky toes. Finally, with her lover still holding tight, she selects a leaf which hangs over a pond of still water and crawls down and places her cloaca against the tip of the leaf. As the eggs begin to emerge, she drags the male upward toward the stem, but as they go, the two frogs together roll the leaf into a kind of funnel, at the same time filling it with a foamy mass of eggs and sperm. They will leave the foam in its funnel, the tip overhanging the pond. As the tadpoles emerge, they fall, one by one, through the carefully aimed hole directly into the water below.

RHINOCEROSES

The Hell You Say!

Rhinoceroses are lousy lovers, at least according to traditional human values. However, it might not be fair to look for tenderness and sensitivity in armor-plated behemoths. The African black rhino is enormous (although not the largest of its kind), standing five feet tall at the shoulder, weighing 3,000 pounds, and sporting two horns, the longest reaching perhaps four feet. They are indeed impressive and powerful beasts. The female may be a bit smaller than the male, but she is just as mean.

Because of the size and temperament of the beasts, their mating can be expected to be a bit rough. But no one would imagine how rough it really is. In fact, rhinocerous courtship is so dangerous that sometimes one of the animals is killed. Zoos are often hesitant to breed them since they don't want to risk losing either one.

The first signal that the fracas is about to begin is the swelling of the female's vulva. As she enters her mating period, her appetite falls off, and she begins to make shrill whistling noises when in the presence of a male. The male responds with deep, heaving sighs. Then they begin to get a bit edgy, tossing their huge heads and trotting about nervously. As the tension continues to build, sooner or later something has to break. It usually happens when the male wheels and charges his lady love. She is far too powerful in her own right to be intimidated by him, though, and she meets his charge head-on. The force of the impact is incredible. Again and again the beasts

Courtship among rhinoceroses is an incredibly violent and dangerous affair. Once the female accepts the male, however, he proves to be very amorous as he climbs farther and farther up her back with each ejaculation.

rush each other, butting and hooking powerfully with their curved horns. The whistles and moans have now given way to great bellows of anger.

An hour later, the fight will grow increasingly savage, and it is now, when they are approaching exhaustion, that one of the animals will likely be injured. It is also about this time that the female will decide whether the male is worthy. If she decides to mate, she becomes no gentler, but instead of counterattacking, she begins to parry his lunges, as if to behave as coquettishly as the moment and her ponderous body will allow. Finally she signals her willingness to cooperate. The fighting is finished, but the worst is not yet over.

She presents her rear to the male and pulls her tail tightly to one side, bringing her swollen vulva into plain view. At this the male changes his entire attitude. He moves around her, rears up, and mounts her from behind. He is less than the perfect lover, however, partly because he lacks finesse and partly because his penis is about two feet long. He thrusts and grunts but has great trouble penetrating her vagina, and when he finally succeeds, he may drive home only the first few inches—at first. After all the trouble he went to to get there, however, he's not about to quit until he's ready. And that may be quite some time. He may remain mounted for over an hour, and he ejaculates about every ten minutes. With each ejaculation, he seems to take a greater interest in his business until finally he has buried the full length of his penis. Worse yet, by now he has *literally* mounted her, and all *four* of his feet are off the ground! The female staggers around, his great ungainly body sprawled across her back in a most unseemly fashion. When he finally withdraws, both animals are utterly exhausted. Sometimes they quickly recover, however, and the whole process is started all over again about an hour later.

She will be pregnant for about sixteen months, and when she gives birth, it, too, is an unusually violent process. When the time comes, she stands heaving, her great head held low and her sides swelling mightily. Finally, the 130-pound calf is popped out of her body, almost like a champagne cork. As it feebly kneels there in its pool of blood and tissue, she assists in expelling the afterbirth by walking backwards, treading on whatever part is already on the ground, and pulling the rest out as she goes.

FISH

And the Total Feminist

It is not an easy job to describe how fish do it because there are many kinds of fish, and they lead different kinds of lives. We can only be sure that they do it in water—which is the excuse W. C. Fields gave for never touching the stuff.

Some fish, such as cod and herring, take the easy way out and simply gather in groups and promiscuously release millions of eggs and sperm directly into the water, leaving the rest to chance. But that's not very interesting.

Other fish have very complex mating practices. The male stickleback (a little fish of European ditches), for example, entices a female into a nest covered by a tunnel of twigs. He then prods her with his snout until she releases her eggs and dutifully fans oxygen-laden water over them until they hatch.

Among the species in which the sexes release eggs and sperm into the water, the chances of fertilization are obviously increased if they are released in close proximity. Since both eggs and sperm are discharged from the "anus" (actually a cloaca), this may mean that the fish should simply get their anuses as close together as possible. Of course, that's not really doing it, but copulation undoubtedly evolved as a way of simply getting the eggs and sperm close together in a hospitable environment. In other species, such as the cottidae, the little mound of flesh around the male's anus may become greatly enlarged and during mating protrude as a kind of penis which he can aim directly at the eggs. In another species, the female's egg duct is protrusible, and she pushes it

The female grunion has burrowed backwards into the tidal sands of a California beach. The male, lying over her exposed head, is releasing his bubbling mass of sperm over the eggs as they emerge into the protective sand. The next high tide will wash the fertilized eggs out to sea.

In this East African mosquito fish, sperm has been passed through grooves in specialized fins to the penis, or gonopodium, which is itself a modified fin. These fish practice true copulation as the darting male swiftly inserts the long organ into the bodies of females.

against the fleshy little mound above the male's anus until it becomes smeared with his sperm. She then pulls it back into her body and releases her eggs past the captured sperm. In the cardinal fishes, the genital mound of the female becomes enlarged, and she sticks it directly into the male's anus to harvest his sperm; in the bitterling, even though the fleshy mound of the female is tremendously elongated, she doesn't stick it into the male. Instead, she slips it into the gill spaces of a freshwater mussel and lays her eggs there where they will be relatively safe, and the male simply pours his sperm over the mussel.

Fortunately for our story, some fish are more intimate in their mating. That is, they really do it. In these species, the "penis" of the male is modified from the rays of the anal fin. Sperm, or packets of sperm, pour into a trough formed by the hindmost rays. They pass from there through a deep groove in another ray to the tip of the longest ray—the actual "penis" or gonopodium. This penis is usually a long, delicate structure which the male is forever seeking to introduce into the anus of some female.

Male guppies are continually darting under females, quickly trying to slip their penises into them. These furtive goosings are probably not true matings, however, since sperm are usually transferred only when the tip of the penis is allowed to remain in the female for a matter of seconds. Once inserted, it may be held in position by a hook and claw located at the tip of the organ. True copulation requires the participation of the female, however; she must first signal her willingness to mate by drifting and swimming in circles before the trembling male.

The male guppy ejaculates only when his penis is fully erect and the base is positioned directly below the female's anus. During copulation, the female may help by secreting fluids which enter the male's penis and "thin" his semen, causing it to move faster. The sperm, thus released, swim easily along the penis to the female. The young will be born alive—tiny, darting specks which will be ignored unless the fish are overcrowded. In this case, their parents will eat them.

Sexual identity in fish can be confusing. For example, many are hermaphroditic, harboring both male and female organs. And some, such as the Mediterranean serranids, can even fertilize their own eggs (externally, in the water). And then other species, such as the labrids, are males when they are young, but as they grow, the ovaries begin to mature, and they become females.

Perhaps the strangest case of all, however, is that of the toothcarp of Mexico and Texas. *All* individuals are females! Males simply do not exist. When they mate, these amazons seek out males of a related species and dupe them into donating their sperm. Those sperm, however, do not actually unite with the eggs and contribute their genes to the offspring— they are only needed to stimulate the eggs to begin their development. Once this is accomplished, the sperm falls uselessly away from the egg, carrying its genes with it. No one knows what happened to the males of this species or how they came to be so totally useless that nature, ever more efficient, ordered them to oblivion, but they are banished forever from this totally feminist world.

HIPPOPOTAMUSES

That's When It Hit the Fan!

Hippopotamuses can behave very strangely toward each other. A father may bite off the head of his son, and a captive adolescent may suddenly gore his mother to death with his huge canine tusks. It's no wonder that the females seek to simplify their domestic life by staying together, consorting with males only to copulate.

These groups of females are usually comprised of forty to fifty individuals which spend their days on river bottoms with only their eyes and nostrils projecting above the waterline, or completely submerged, occasionally rising to the surface like great corpulent bubbles.

They are remarkable beasts in a number of ways. For example, should a hippopotamus find itself out of the water for very long, it will mysteriously begin to sweat blood. Actually, the "blood" is only a very dark red and oily kind of protective sweat, but it certainly gives one quite a start! They are also surprisingly agile, able to climb steep bluffs when they emerge at night to graze on the greenery around the river.

Each animal stakes out a small and narrow territory and marks the area in a most dramatic fashion. As it defecates and urinates at the same time, it whirls its tail like a propellor, scattering the mess in all directions but rendering its land unmistakably its own.

We, with our tiny noses, tend to think of hippopotamuses

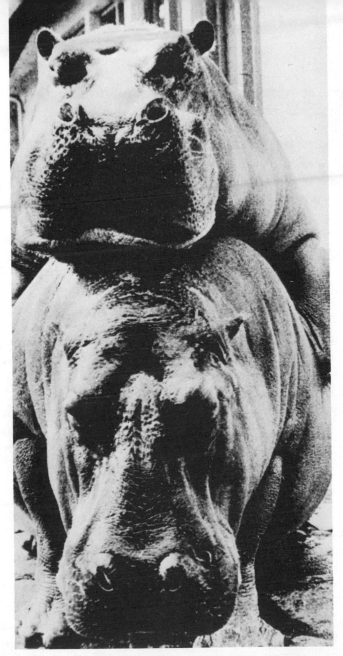

Hippopotamuses usually copulate in the water. Since the male's penis points downward, he uses his buoyancy to lift his bulk over the female's body. Their behavior before copulation is most unusual. The males are always dangerous, and females allow them around only to breed.

as ugly, but they obviously find each other quite attractive as demonstrated by their willingness to copulate, whether in captivity or in the wild. They do, however, have the good grace to do it under cover of night.

An observer floating along an African river might be haunted by the stories of overturned boats and men bitten cleanly in two, but his fears might disappear as he sees the great beasts at "play," rearing and bucking around each other and splashing mightily in the shallow water. Actually, though, his insecurities may reappear when he learns that they are not playing at all but are engaged in precopulatory behavior. The irregular monthly menstrual cycle of the female guarantees that some individuals are likely to be breeding at any time, so the sight is a common one.

Mating usually takes place in the water with the male's great weight partially relieved by his buoyance. His penis points backwards, a very effective arrangement for scattering urine but a bit inconvenient for copulating. He mounts the female from the rear in the classical fashion (if either of those latter words can be applied to hippopotamuses). When erect, his penis points a bit forward and downward, so he must position himself in a rather high stance. With his great head lying along the female's back, he probes and pushes until he enters her. Then he gives several heaving thrusts and ejaculates a thick, heavy semen deep into her vagina. The semen tends to coagulate quickly, so little of it is lost by running back out of the mucus-lined vagina into the water.

Her period of pregnancy lasts eight months, a surprisingly short time for such a large animal. When her time comes, she chases away any of the dangerous males who happen to be lingering nearby. She then settles into the water for her ordeal. Her great vulva begins to swell rhythmically and protrude grotesquely. Suddenly it opens, and the infant is literally spewed out and propelled directly to the muddy bottom. It immediately bobs to the surface, only to be pushed under again by the mother. Then, with the mother lying on her side on the river bottom, the infant begins to nurse. When both animals rise to the surface for air as they must frequently do, the infant gets its first glimpses of another, and more threatening, kind of world.

PENGUINS

Such a Nice Couple

Probably no creature suffers so much in the name of parenthood as the hapless penguin. Because of this, and quirks in the human value system, its family life is greatly admired by those people who know anything about it. Of course, we like penguins for other reasons, too, such as their formal dress, their waddle, their upright postures, and their trusting nature on land, which, in the last century, allowed them to be driven by the thousands to the boiling pots.

The emperor penguin is the largest of all, standing four feet tall and weighing up to a hundred pounds. When the Antarctic winter begins in March, the males clamber ashore just as the sea begins to freeze over. Soon thousands of them will be milling around, anxiously glancing seaward as they wait for the arrival of the females. Finally, at about the end of April, the females arrive, all of them a bit travel-worn. By now the frozen surface has advanced so far seaward that the females have had to clamber mile after mile over the icy wasteland, driven by some compelling force, anxiously searching out the males. But in spite of their desire, upon their arrival there is no orgy. In fact, most of the females are already "married," and they begin waddling through the throng looking for their mates. When a female approaches a likely prospect (they all look alike), she sizes him up carefully and sings a little cadence. If his response doesn't sound right, off she goes to keep looking until she finds her husband. When a pair is reunited, they stand breast to breast, heads

This male king penguin is attempting to mount the female from the rear. The female obligingly leans forward, about to lie on her belly, but their rounded bodies make balancing very difficult, and if he proves inept she may get up and waddle away.

thrown back, singing loudly, with their outstretched flippers atremble.

If a bird has been unable to find its mate, it will finally accept a new spouse. But should the old partner show up a day or two later, the new engagement will be immediately broken and the latecomer welcomed.

The pair will spend the next few days walking around side by side, occasionally touching each other with their flippers and bills. No matter how attentive they are during the day, at night their main interest is survival as winds reach one hundred miles per hour and the temperature plummets to seventy degrees below zero. Then they huddle together in great masses or circle in an endless parade through the night.

Two weeks after a pair is formed, the union is consummated. The male's intentions first become obvious when he lays his head across her stomach. She gets the message and they leave the colony, trundling over the icy ridges until they find a secluded place. Once they are alone, the female lowers herself onto the ice, facedown, and the male mounts her from the rear. Because their bodies are so round, however, he has trouble staying astride her, although he grasps her beak in his and balances as best he can with his flippers. When he feels securely mounted and has gained his composure, he pushes his tail downward and presses his cloaca against hers, which is raised in anticipation. He is not very good at this, and he may lose his balance and tumble off her back a few times. If this happens too often she becomes impatient, gets up, and waddles off in disgust. If they were as formal as they look, he would, of course, need to undergo counseling. But as it is, he simply begins the entire ceremony anew.

When he doesn't bumble and manages to press his cloaca firmly against hers, he quickly ejaculates, and the entire process will not have taken over three minutes. Neither one will mate again that year after they return to the hubbub of the colony.

The single egg is incubated between the feet and a bare warm place on the belly. Soon after it is laid, the female turns the responsibility for it over to the male and goes off for a snack. She will be gone about two months (since the sea may, by then, be fifty miles away). When she returns, fat and with a stomach full of undigested fish for the coming hatchlings, the gaunt male can finally give up his vigil and go find his first meal in four months.

TURTLES

Stick It Right Out There!

Turtles, terrapins, and tortoises all have the same problem: their shells. Put simply, it's hard for a male with a flat belly to mount a female with a rounded back. Even when everything goes well, it gives one the impression that it won't. After all, here we have two stupid, brutish, and clumsy beasts trying to perform an act which obviously requires a certain degree of deftness and dexterity. So how do they do it?

Doing it, for turtles, involves the male getting his penis into the female's cloaca, just under her tail. His penis is normally tucked away in the front part of his own cloaca, but when he becomes sexually excited, it begins to engorge with blood until it everts from its pouch by turning inside out. A turtle has a long penis because of the shell problem—it has to reach a long way. It carries no urine, only sperm—which flows along a deep groove in the warty, tapering organ.

Now the male has problems besides those relating to simple mechanics. For example, the female herself may be coy or just downright unwilling, and in such cases he must try to convince her. At first he may take the indirect route: courting. He approaches her head-on, and if she is not repelled, they stand there, their heads nodding and bobbing, perhaps for hours at a time, the male nibbling rakishly at her toes. In the painted turtle, the male has special elongated "fingernails" with which he reaches over and strokes the female's cheek. She likes that. If they are in the water, he may swim around

This male South American forest tortoise has clumsily mounted his mate. He must balance precariously atop her rounded back. His grooved penis is long, pointed, and covered with warts, but in spite of its beauty, successful copulation requires her complete cooperation.

in front of her, showing off his brightly colored forefeet. She seems to like that, too.

Among mud turtles, the male swims slowly toward the female until their heads almost touch. Then they begin to extend and retract their heads in a most peculiar way. At the same time, their forelegs tremble and quiver, and in some species, they then begin a ritualistic "swimming dance." Even immature mud turtles of these species greet each other with the quiver and dance.

In the green sea turtle, the males are sexually aggressive throughout the breeding season. But as luck would have it, not all the females are receptive. If she is not in the mood, she will first swim swiftly away from a suitor. If he pursues, she may even crawl ashore to get rid of him (it always works). If he catches her in the water, she may pointedly press her hind legs together, much as a woman might cross her legs. If he doesn't take the hint, she may simply bite him. Sometimes, though, she signals her unwillingness by "standing" upright in the water, her belly turned toward him and her four limbs extended. This posture seems to put him off, and he usually swims away.

Male sea turtles are not easily put off, though. In the breeding season, they may try to mount anything in the water. Amorous males have even attempted to mount curious scuba divers, much to the divers' chagrin. The divers finally found that they could float spread-eagle in the water in the manner of unwilling females and save themselves further indignity. Amorous males sometimes even try to mount each other. However, if a male suddenly finds himself about to be ravaged by another male, he uses none of the female's signals but lunges and thrashes about and finally turns upside down, at which point his nearsighted suitor moves on.

Of course, not all females are unwilling. When one is receptive, the male mounts her from behind (except for one possible exception—in conservative Australia, where they may copulate belly to belly). In many species, the undersurface of the male is slightly concave to enable him to ride the female's shell a bit more closely. In the sea turtle, the male hooks his thumbnails over the front of the female's shell (sometimes slightly cutting her neck) and maneuvers the back of his shell downward until he is able to introduce his penis into her cloaca. Once he has entered her, they may remain joined for as long as a day.

Box terrapins have a little more trouble. The tail of the female is very short and her cloaca doesn't protrude very far, so it's hard to get under. In order to reach it, the male has to sit almost vertically, his forearms pawing uselessly at the air as he holds himself in place by gripping the rear of the female's breastplate with his back claws. The female helps him with his balancing act a bit by bracing his ankles with her hind legs.

Male land tortoises become particularly amorous during their breeding season and may simply refuse to take no for an answer. But they require the female's cooperation. In order for them to mate, the female must protrude her rear end as far out of the shell as possible. If a female should draw her rear in uncooperatively, the male simply marches around to her front and snaps at her head, which she quickly jerks in. But in the breeding season, tortoises are fat, so when her head goes in, her tail goes out. He then returns to her protruding rear and mounts her at his leisure.

In some tortoises, the male uses his knotty tail to keep the female aroused during copulation. While they are at it, he continually slaps at her cloaca with his tail, stimulating her and causing her to extrude her rear even further.

Male turtles of almost all species make some sort of noise while copulating. Charles Darwin described the roars reverberating across the islands during the mating of the Galápagos tortoises. Smaller species, though, sound more like cats yowling or babies crying. The sounds carry well, and females who are mounted by a wailing male often stretch out their necks and jerk their heads around as if trying to see who's making all the racket.

DOLPHINS

A Swiveling Penis

We've all heard how intelligent porpoises are from every-
one from the *Reader's Digest* crowd to environmentalists
seeking to save them from tuna fishermen. We value intelligence, and so many of us are curious about their life histories,
including their love lives. And it turns out that their love
play, by human terms, is quite sensuous if not particularly
imaginative. But their lack of imagination can be forgiven
since a great deal of improvisation should not be expected in
an animal built like a torpedo.

Dolphins and whales are both distant cousins of land mammals. Apparently they are descended from land creatures
which returned to the sea. In adapting to the demands of the
water, however, they sacrificed much. For example, they can
no longer clasp each other tightly with their stiffened fins,
and their rough, protective skin has undoubtedly lost much
of its sensitivity.

Their foreplay, however, can be remarkably gentle. In the
bottle-nosed dolphin, the sexes swim around each other, rubbing and touching one another with their snouts and fins. At
such times, the female may nuzzle the male and rub her head
along his belly. Then she may break away and swim under
him upside down, genitals in full view.

The vulva of the female is tucked into a slit under her body,
which may or may not also harbor the anus. The opening to
the vagina is quite small, probably to keep seawater out and
sperm in, but she has good control over the muscles surround-

ing it so she is able to admit the male's penis at the proper time.

His penis is a long, muscular organ lying curled deep within his body behind his testicles until it is erected. Then its peculiar structure becomes apparent. The tip, which is the only part inserted into the female, seems to be attached to the rest of the penis by a kind of swivel so that it is free to swing and rotate independently of the shaft. And unlike the human penis which seems to have a mind of its own, it is under the strict control of its owner.

Captive dolphins tend to masturbate freely, even when in the presence of females. One woman keeper made news by occasionally masturbating her male charges when they approached her with an erection. The extrusion of the penis from where it lies coiled within the body, however, does not necessarily lead to sexual activities. Since the penis is flexible and sensitive, it is ideal for exploring things along the bottom of murky water, and male porpoises have been seen poking and probing their way around the bottom of a new tank in a most surprising manner.

As the foreplay of dolphins becomes more serious, the innocent rubbing may give way to other behavior. For example, the male may swim around with his back fin firmly and rather pointedly inserted into the female's vagina. As their excitement grows, both animals may slap the surface and leap clear of the water before they return to courting. The male may then gently bite the back fin of the female, especially if she continues to present it to him. She seems to be stimulated by his bites, and after a nibble she swims quickly away upside down, flashing her genitals as she goes. Jealous females display themselves in the same way to lure males away from other females with whom the males are already mating. The male, for his part, displays himself by standing vertically in the water with a full erection, or he may swim upside down under a female, his penis cutting through the water. When a male with an erection approaches an unwilling female, she may roll over, presenting her back to him, or with her flipper she may even direct blows at his head or, worse, his penis.

When the foreplay is finished, the bull grasps the cow as best he can with his flippers, holding her belly against his. With his mouth agape at this time and emitting subsonic squeaks and squeals, he thrusts his penis against her until it enters her vagina, whereupon she exhales noisily through her blowhole. He then makes a few pelvic thrusts, but there is no change in his behavior to indicate that he has ejaculated. He simply lets her go and swims away.

In the Ganges dolphin, a freshwater species, copulation is a bit more spectacular. The pair plunge beneath the water, turn, and swim upward and toward each other, breaking free of the water just as they collide, belly to belly, hugging each other clumsily with their flippers, half their bodies out of the water and their tails beating furiously. The male penetrates the female in this instant, and they fall back onto their sides to roll together briefly over the surface of the water. Then they separate and swim swiftly away.

When a female of an ocean species gives birth, she will be surrounded by others who swim around her, guarding her from any sharks which might be attracted by the blood. The newborn dolphin may have to be aided to the surface for his first gasp of air. Then he will return to nurse from the nipples along his mother's underbelly, showing that he is really no fish at all.

LIZARDS

Make Him Feel, You Know... Dominant!

Among lizards, males are dominant. In fact, if the females are not subordinate and passive, mating cannot occur at all.

The males first show their dominant nature in the breeding season, when they set up territories and vigorously drive away all other males. A male on his territory may first threaten an intruder by performing rapid pushups with his forelegs. If that doesn't work, he may do four-legged pushups—a feat sure to impress anyone. Should that fail, he will attack. Usually the intruder loses; he is not fighting on his home soil.

Should a female enter a male's territory, he employs a different approach. He stalks stiffly up to her, his head held high. His brightly colored throat is now in full view, and he is the picture of confidence. His confidence may be misplaced, however, because the lady may suddenly turn and hightail it.

But if she is ready to copulate, she will greet his approach by stamping the ground rapidly with her forefeet and sending waves undulating along her tail. These may actually be "submissive" signals, since defeated males perform the same movements in the presence of their conquerors. But even if the female is ready, she is also coy. She may now turn and run from him—but not too fast. The male is easily able to keep up with her, and he scurries along behind her with his snout touching her tail. Then, even while they are running, he

The male chameleon signals his readiness by very stereotyped movements. If the female accepts him, he will lie across her so that one of his two barbed penises can enter her cloaca. He must withdraw very slowly. If she is startled she may scurry away, dragging him ignominiously along by his penis.

begins to lick her tail. She slows down. He licks higher and higher, and she runs slower and slower, until she stops completely. At this point he may shove her along a bit as if he wants her to run, but if she can't be budged, he changes his behavior and bites her flank. He holds her in his jaws for a moment and then moves his grip up to her neck, the first step in mating for many lizards. (Males of some species, though, prefer their sex a bit more conventional and omit the licking.)

The female is not necessarily overpowered by the male because in many species she is easily as large and strong as he is and *could* fight it out with him toe to toe. If she wishes to mate, however, she must acquiesce, acknowledging his dominance. He will not mate with a brawling female.

Not all species follow this precise sequence, of course. Visual signals such as pushups are more important to those lizards who hunt by day, and signaling by scent and touch is more common in species which are active at night. Among the day hunters, many males signal by expanding and retracting their long, thin throat pouches. It is interesting that some of those who do pushups continue to do them even while they are trying to copulate—a bizarre example of lizard machismo.

In any case, once a male has the female by the neck, he usually only does a few more quick pushups and then drags her along for a way. The male side-blotched lizard may jerk the female's head from side to side, causing her tail to sail back and forth like a windshield wiper. But finally he will place one hind leg over her back and begin to stroke her gently across the hips with it. At the same time, he begins to push the base of his tail against hers. She responds by raising her hips and tail, causing the lips of her cloaca to spread apart. The male immediately slips his tail under hers and inserts his penis.

Actually, he inserts one of his penises, since he has two. They are spiny fingers which lie side by side in the same arrangement found in snakes and which are erected by turning inside out from the wall of his cloaca.

Once he has inserted his barbed organ into the female, he may thrust against her several times before one final shove which sends his penis in as deep as possible. Then his tail jerks spasmodically several times in what is probably an ejaculation. Occasionally the female begins to move away before he has withdrawn his penis, and since it is held fast by barbs, he must hobble desperately along behind her. However, if he should jerk his head rapidly from side to side in a "no" fashion, she immediately stops. Then he quickly uncouples and she goes on her way, leaving him standing there squinting after her with his penis hanging out.

BACTERIA

Primitive Sex

Some bacteria do it, but they don't have to. After all, any bacteria can multiply by dividing; simply pinching in two. This is a very efficient and rapid process, as you know from your experience with the spread of infections. But it obviously doesn't permit genetic mixing, and mixing of genes is what sexual reproduction is all about. Mixing produces instant genetic variation, and the more variation there is in any population, the more likely some individuals are to be able to survive, whatever happens. And the mixing somehow seems to produce healthier, more robust offspring.

It's probably not quite right to say that bacteria have sex because, actually, they don't even have sexes. But different "types" exist within populations, and certain people have seen fit to call these "male" and "female" (ignoring the fact that there may be more than two types, though it is true that each type can only do it with certain other types).

In any case, in bacterial copulation, one partner "donates" its genes to the other and is therefore called the male. When he finds himself near an individual of another type (called a female), he slowly protrudes a part of his body to form a long, slender channel which reaches over and penetrates her body wall. Then through the tubes he ejects a certain kind of genetic material called the F factor (F stands for fertility, in spite of what your first guess was). Males have the F factor; females lack it.

The F factor is actually a circular chromosome, and it must

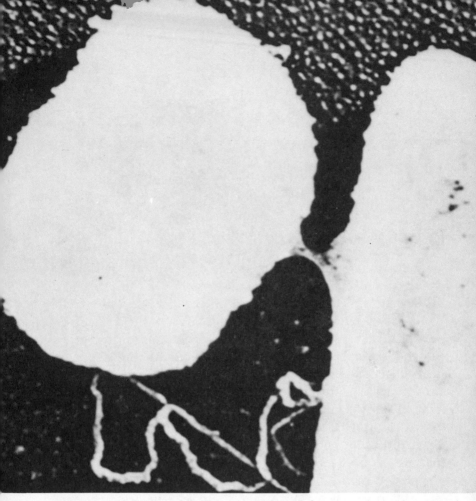

Here, genetic material is being transferred from one bacterium to another across the conjugation bridge connecting them.

break and straighten out in order to wend its way through the long tube connecting the two tiny beasts. But bacterial sex is a delicate process, and often the mating is interrupted (we probably shouldn't refer to the penis being broken) before all the F factor has been transferred. Since any amount could have been transferred before the break, genetic variation in the population is increased as different individuals end up with different amounts of the F factor.

When all goes well, the entire process only takes an hour or so. If the pair should stay joined for that long, all the F factor will be transferred and the female will become a full-fledged male. The old male, however, does not revert to becoming a female. He is done. His only chore is finished, and he dies without further ado. The new male must live on, however, until he too pierces the wall of another animal, giving his life but seeing that his kinds of genes continue in the world's population of bacteria.

This is undoubtedly the simplest sort of sexual reproduction. Perhaps it was the very first kind to arise after life evolved. Nevertheless, it does seem a bit peculiar to see sex transferred as some sort of infection, a kind of bizarre genetic disease.

BIRDS

A Different Kind of Kiss

It is easy to describe how birds actually *do* it. It's the preliminaries that are the problem. The reason is that with about 9,000 species of birds on the earth, there are about 9,000 ways to initiate things. For example, the "strategies" may be entirely different in species in which the sexes look alike and those in which they don't. Drab little male sparrows may have difficulty in convincing drab little female sparrows to mate with them. The birds may have to consort with each other some time before each decides the other is really of the opposite sex and would make an adequate mate. Once they mate, however, their drabness pays off as they inconspicuously take turns sitting on the nest and feeding the young.

At the other end of the spectrum are the peacocks and the birds of paradise. In these kinds of species, the females are drab, all right, but the males are gorgeous. They turn on their females quickly by barraging their instinctive centers with bold markings. They won't be found sitting on any nest, however. And actually, the female might well be pleased that the male leaves her after mating. The last thing she needs is some flamboyant male hanging around and attracting predators to her or the nest. Besides, in these species, she is easily able to find enough food for the young by herself, thus releasing him from those mundane household chores. So after he mates with her, he disappears into the sunset looking for new conquests. In fact, this makes sense because since he is so conspicuous, his best bet is to mate as often as possible

Birds are seemingly not equipped to do it at all since they lack penises and vaginas. It usually requires a delicate balancing act, as we see in gangly shorebirds. The act itself may take only a split second as the cloacas are pressed together and the male ejaculates.

before the inevitable predator spots him. His motto is, "Live fast, love hard, die young, and leave a beautiful memory." She, on the other hand, lives quietly and mates only once a season but she may survive several seasons. Evolutionists consider the mating system of these birds "advanced" over the monogamy of the sparrows.

Perhaps the most advanced system, however, is that of the bowerbirds. Paradoxically, these males have become drab again. After all, it's dangerous to be conspicuous. It turns out, however, that they *are* attention getters, but they attract the attention to something else. What they do is build a "bower" which may be a tunnel of grass, a cleared area, or a small pole, but in any case, it is garishly decorated with all sorts of doodads—paper, leaves, cellophane, bits of glass, feathers, small bones, shells, yarn, or anything else that might appeal to some female. (She often shows a distinct preference for blue.) When the bower is finished, the male stands near his creation and noisily calls to any females in the area. When one shows up, if his nest is impressive enough, the materialistic lady will mate with him. Actually, she has no use for a bower, but it indicates his general health and vigor. If she mates with him, then her offspring will receive the genes of a strong, healthy father. She, of course, doesn't think about all this, but the "successful" females are those who behave as if they do.

When a female has made her decision (the choice is almost always hers—that's why the males are driven to such extremes of salesmanship), copulating is rather simple. The male usually walks up on the tail of a crouching female, balances himself by grabbing her neck feathers and fluttering his wings, and tucks his tail under hers until the lips of their anuses (actually cloacas) are pressed firmly together. He ejaculates at once, squirting sperm with some force into her. If he is inept or dallies too long, she may shake him off and walk away. Birds are often prey animals, after all, and she can't remain in a compromising position while some awkward male gets his act together.

The male of some birds, such as ducks and ostriches, has a penis which, when he is aroused, protrudes stiffly from his anus. It is not a "tube" but a fleshy rod with a deep groove through which the sperm rushes. This condition is obviously more advanced than the cloacal kiss—and is a surer method of getting the sperm into the female.

Male ducks of some species become veritable sex maniacs in the spring. Although a male will stay "married" to his female (at least for the season), he will immediately try to force himself on any female he sees. "Husbands" may not protect their females from such advances, but if his mate is being chased by a strange male, he may go along to make sure he doesn't lose her in the commotion. For her part, she would apparently rather die than be unfaithful and may spend the whole day hiding in the grass to avoid the interloper.

Geese, on the other hand, handle such problems differently and may readily form triangles. In fact, if two males establish homosexual bonds, they may prefer each other's company to any female's. Sometimes, however, when they are courting, a female may suddenly interpose herself between them and be quickly fertilized. She is accepted quite readily by them, and the three can be seen together some weeks later attending to several tiny goslings, a happy—if a bit kinky—family.

CAMELS

One Hump or Two?

Camels are nasty animals. Of course, both the single-humped Arabian and the double-humped Bactrian have been important to human "progress" in the drier areas of the world. After all, they are big and strong and are well adapted for desert living with their double rows of eyelashes, hairy ears, and big feet—qualities which help them survive and make them ugly at the same time. Although their humps do not store water, but fat, they can go without drinking for as long as two weeks. Camels also have the distinction of being the first animals to use IUDs. For years, camel drivers placed apricot pits in the uteri of their females so that they would not become pregnant on long caravan marches. (One might well wonder how the practice got started. There must be a better place to store apricot pits.)

The more one deals with camels, the weirder they (the camels) seem to get. For example, no matter how much love and attention have gone into their rearing, when they are adults they will invariably try to bite their keepers. And the foulness of their temperament is matched only by their smell. In addition, they can muster a powerful death wish. For example, in zoos, if they are moved to a new paddock, they may all simply lie down and die. Autopsies often show no organic disorder whatsoever.

And to top it off, they are lousy parents. About the best that can be said of the males is that they generally don't attack their offspring. Even the females seem to have very

Sex among camels is startling even to people familiar with their habits. The male's mating behavior is particularly bizarre, if not grotesque. The female must usually be subdued before she will allow the male to mate with her.

weak parental instincts. They often don't even know how to nurse their young, and the introductions have to be made by human handlers.

Their love life is no more appealing than anything else about them. To begin with, it's rowdy. When the female comes into heat, the pair may battle furiously, biting, kicking, and hitting with their heads. In the midst of the fray, the male may suddenly break away and just stand there, grinding his teeth, his throat grotesquely swollen, while he somehow manages to make deep and most unpleasant bubbling sounds. But the clincher, for a human observer, comes when he opens his mouth and extrudes his soft palate, a grotesque mound of pink flesh which protrudes from between his lips. At this point, even the most ardent observer is likely to walk away, waving his arms and muttering, "All right, that's it, that's *it*!"

Anyone who stays around, though, will see the male continue to circle the female, biting at her shoulders, pushing his neck down on hers, and swinging at her forelegs with his head. When she is finally exhausted and stops fighting, he rears up and mounts her from behind. But he is not trying to copulate. He is trying to force her to the ground. When she gives in, she drops to her knees in the familiar resting position for camels, or she will sit down like a dog. Either way is fine with him. He then places his forelegs on either side of her and sits down behind her, his forelegs extended. In this position, he erects his long, fibrous penis and probes for her vagina. When he enters her, he manages to thrust, but not very vigorously, for about fifteen minutes. There is nothing in his behavior to indicate his ejaculation; he just stops.

When he withdraws and clumsily lurches to his feet, he moves away and stands blankly or begins to feed. But he will show no further attachment at all for his mate. Nor she for him. Perhaps, though, this is the easiest part to understand.

SKUNKS

Attempted Rape

Skunks are somehow able to stand each other well enough to mate in the classical fashion of small carnivores. This means there is quite a bit of rather undignified scuffling which ends with the male holding the female by the neck while they copulate.

When the female comes into heat, her vulva swells slightly, and she begins to discharge mucus. The male is strongly attracted to her at such times and follows her about, sniffing and licking frantically at her genitals. When she moves away, he follows—sometimes by walking backwards for some reason. She seems to weary of his constant attention and eventually attacks him. But the fights are not as ferocious as they seem because neither animal seriously inflicts damage—at least not at first.

Then, after they have fought several times, the male rushes the female and grabs her by the scruff of the neck. But he acts as if he's not quite sure of the next step. He may hold her down for ten minutes or so without making any attempt to copulate. In fact, as soon as she recovers from the rough treatment, she may get up and drag him around the area. If she finds food, she may casually begin to eat with the male still hanging on. If this happens, he will give up his hold to see what she has found. Then he grabs her again and drags her away from the distraction.

He may now begin to get serious by clasping her around the waist and mounting her from behind, his penis fully erect

—such as it is. It is only about an inch and a half long and about as thick as a pencil lead. He still can't enter her, though, unless she cooperates. So he stimulates her in a unique way; he begins to scratch vigorously at her vulva with his rear leg. At first she ignores his efforts, but soon he seems to cross her erotic threshold, and she straightens her rear legs and moves her broad tail to one side. Sometimes she drops down on her forepaws and places her chin on the ground.

He now begins to thrust rapidly until his tiny penis enters her swollen vagina. At this point, he throws her onto her side and pushes with his uppermost foot against the back of her upper leg, making her vagina even more accessible, and continues to thrust. As soon as he ejaculates, he withdraws his penis and frantically begins to groom and lick at his own genitals. She shows no such obsession with personal hygiene and simply walks away.

He may copulate with several females in a very brief time since the willingness of one seems to be contagious, giving them all the idea. (There may be several females around since they may all have shared their sleeping quarters through the long winter months.)

As the female's period of heat ends, she becomes less receptive to the male until finally she begins to fight him off by falling onto her back and biting and kicking. His ardor has not waned, however, and he may attempt to rape her, grabbing her by the neck again and again and attempting to turn her over. But he is almost never successful. In fact, if he persists, she may give him some nasty bites around his head and face. After she has once roughed him up, he becomes cowed and only seems interested in avoiding her. She may express her disdain for him in no uncertain terms by taking her bedding out of the nest, carrying it away and sleeping in another place. Later, however, when she cools down, she moves back in, apparently willing to forgive and forget. But he hasn't forgotten. Things are never quite the same between them again. At least not that year.

SELECTED
SOURCE MATERIAL

This is a partial list of the more pertinent literature used in the preparation of this book.

Anderson, H. T. *Biology of Marine Mammals* (Chapter 8, by R. J. Harrison). Academic Press, 1969.
whales, porpoises

Aronson, L. "The physiology of fishes," *Behavior*, Vol. II, ed. M. Brown. Academic Press, 1957.
fish

Bellairs, A. *Life of Reptiles*, Vol. II. Universe Books, 1970.
turtles, snakes, alligators, crocodiles

Bellairs, A. "Reproduction in lizards and snakes," *New Biology*. Penguin Books, 1959.
lizards, snakes

Biggers, J. D. "Reproduction in male marsupials," *Comparative Biology of Reproduction in Mammals*, ed. I. W. Rowlands. Academic Press, 1966.
opossums, kangaroos

Bishop, S. *Handbook of Salamanders*. Cornell University Press, 1943.
salamanders

Booth, J., and J. Peters. "Behavioural studies on the green turtle (*Chelonia mydas*) in the sea." *Animal Behavior*, Vol. 20 (1972), pp. 802–812.
turtles

Buddenbrock, W. B. *The Love-Life of Animals*. T. Y. Crowell, 1958.
deep-sea fish, dragonflies, seahorses, lizards, birds, frogs, rabbits, spiders, sharks, earthworms
Budker, Paul. *The Life of Sharks*. Columbia University Press, 1971.
sharks
Burns, E. *Sex Life of Wild Animals: A North American Study*. Rinehart, 1953.
bats, dogs, fish, birds, porcupines, rats, whales, dragonflies
Davey, K. *Reproduction in the Insects*. W. H. Freeman, 1965.
dragonflies, bees, bedbugs, praying mantises
DeRopp, R. *Sex Energy*. Delacorte Press, 1969.
bedbugs, spiders, praying mantises, honeybees, earthworms, snails, octopuses, salamanders, frogs, birds, seahorses, sharks, lizards, snakes, crocodiles, turtles, platypuses, gorillas, dolphins, whales, dogs, cats
Dewsbury, D. "Patterns of copulatory behavior in male mammals." *Quarterly Review of Biology*, Vol. 47 (1972), pp. 1–33.
porcupines, bats, chimpanzees, gorillas, dolphins, horses, dogs, pigs, camels, cats
Engelmann, E. *Physiology of Insect Reproduction*. Pergamon Press, 1970.
bedbugs, praying mantises
Evans, G. Y. *Communication in the Animal World*. T. Y. Crowell, 1968.
fish
Ewer, R. *Ethology of Mammals*. Plenum, 1968.
elephants, kangaroos, rhinoceroses, porcupines, dogs, skunks, cats, camels, horses
Ferguson, G. "Mating behaviour of the side-blotched lizards of the genus *Uta* (*sauria: Iguanididae*)." *Animal Behavior*, Vol. 18 (1970), pp. 65–75.
lizards
Ford, W. S., ed. *Bioluminescence, Pigments and Poisons*, Vol. III. Academic Press, 1969.
sharks
Fraser, A. *Reproductive Behavior in Ungulates*. Academic Press, 1968.
horses, camels, cattle

Galtsoff, P. S. "Physiology of reproduction in molluscs." *American Zoologist*, Vol. I (1961), pp. 273–289.
octopuses, snails

Goin, C. "Amphibians, pioneers of terrestrial breeding habits." *Smithsonian Reptile Forum*, 1960.
frogs, salamanders

Goodall, Jane van Lawick. "Some aspects of reproductive behavior in a group of wild chimpanzees, *Pan troglodytes schweinfurthi*, at the Gombe Stream chimpanzee reserve, Tanzania, East Africa." *Journal of Reproductive Fertility*, Suppl. 6 (1969), pp. 353–355.
chimpanzees

Hafez, E., ed. *Reproduction and Breeding Techniques for Laboratory Animals.* Lea and Febiger, 1970.
cats, dogs, rabbits, opossums, bats

Hafez, E. H. E. *Comparative Reproduction of Non-human Primates.* Charles C. Thomas, 1971.
chimpanzees, gorillas

Hoven, T. *Physiology of Vertebrates.* W. B. Saunders, 1968.
fish, frogs, snakes, birds

Jameson, D. "Life history and phylogeny in the salientians." *Systematic Zoology*, Vol. 6 (1957), pp. 75–77.
frogs, salamanders

Jameson, D. "The population dynamics of the cliff frog *Syrrhophus marnocki.*" *The American Midland Naturalist*, Vol. 54 (1955), pp. 350–360.
frogs

Johns, J. *The Mating Game: Sex, Love and Courtship in the Zoo.* St. Martin's Press, 1970.
penguins, rhinoceroses, hippopotamuses, turtles, camels, kangaroos, birds

Legge, R. "Mating Behaviour of American Alligators at Manchester Zoo." *International Zoological Yearbook*, Vol. VII.
alligators

Loxton, R. "Family life of the mantis." *Animals*, Sept. 1969.
praying mantises

Michelmore, S. *Sexual Reproduction.* Natural History Press, 1964.
lobsters, earthworms, snails, insects, spiders, octopuses

Minton, S., and M. Minton. *Giant Reptiles*. Scribner's, 1973.
crocodiles, snakes

Pankes, A., ed. *Physiology of Reproduction*. Longman, 1960.
horses, opossums, pigs, porpoises, gorillas

Perry, R. *The Unknown Ocean*. Taplinger, 1972.
deep-sea fish

Pilleri, G. "Observations on the copulatory behavior of the Gangetic dolphin *Platanista gangetica*," *Investigations on Cetacea*, Vol. III, Part I, 1971.
porpoises

Rabb, T. "Evolutionary aspects of the reproductive behavior of frogs," *Evolutionary Biology of the Anurans*, ed. J. L. Vial. University of Missouri Press, 1973.
frogs

Reed, C. "The copulatory behavior of small mammals." *Journal of Comparative Psychology*, Vol. 39 (1967), pp. 185–206.
platypuses, bats, porcupines, rabbits, opossums

Reynolds, H. "The act of copulation in the opossum (*Didelphus virginianus virginianus*)." University of California Publications in Zoology, Vol. 52 (1953), pp. 230–240.
opossums

Rockstein, M., ed. *Physiology of Insecta*, Vol. I (Chapter 2, by Jan de Wil de Hoof). Academic Press, 1973.
honeybees

Rovner, J. "Copulation in the Lycosid spider (*Lycosa rabida walckenaer*): a quantitative study." *Animal Behaviour*, Vol. 20 (1972), pp. 133–138.
spiders

Schaller, G. *The Mountain Gorilla*. University of Chicago Press, 1963.
gorillas

Schultz, A. *Life of Primates*. Weidenfeld and Nicolson, 1969.
chimpanzees

"Sexual maturity and first recorded copulation of a 16-month male porcupine, *Erethizon dorsatum dorsatum*." General notes, *Journal of Mammalogy*, Vol. 33 (1952), pp. 239–242.
porcupines

Short, R. V. "Species differences," *Reproductive Patterns*, Book 4, eds. C. Austin and R. V. Short. Cambridge University Press, 1972.
opossums, kangaroos, elephants, horses

Slaughter, B., and D. Walton, eds. *A Chiropteran Biology Symposium*. SMU Press, 1970.
bats

Street, P. *Animal Reproduction*. David & Charles, 1974.
fish, toads, crocodiles, dragonflies, penguins, frogs, earthworms, bats, bees, elephants, hippopotamuses, kangaroos, lobsters, mantids, octopuses, opossums, sharks, ticks, whales

Usinger, R. L. "Monograph of Cimicidae." Entomological Society of America, Proceedings, 1961.
bedbugs

Wendt, H. *Sex Life of the Animals*. Simon & Schuster, 1965.
earthworms, snails, octopuses, spiders, ticks, mites, praying mantises, bees, fish, sharks, deep-sea fish, seahorses, frogs, salamanders, lizards, snakes, crocodiles, turtles, porcupines, cats, platypuses, opossums, kangaroos, whales, bats, chimpanzees

Wight, A. "Reproduction in Eastern Skunk, *Mephitis mephitis nigra*." *Journal of Mammalogy*, Vol. 12 (1931), pp. 40–48.
skunks

Wodinsky, J. "Ventilation rate and copulation in *Octopus vulgaris*." *Marine Biology*, Vol. 20 (1973), pp. 154–164.
octopuses

INDEX

About the Author

Robert A. Wallace is the author of *The Genesis Factor*, a popular book on sociobiology published by William Morrow in 1979, as well as a number of very successful textbooks on biology and animal behavior. He has been a zoologist for twenty years, having received his B.A. in biology and fine arts in 1960, his M.A. from Vanderbilt University in 1965, and his Ph.D. from the University of Texas at Austin in 1969. His research has been concerned with social systems of animals, particularly the island birds of the West Indies and the Indian Ocean. He has taught in departments of biology, psychology, and anthropology in several universities in the United States and Europe. His interests include sailing, painting, horseback riding, and long-distance running. He has worked as a musician, basketball and track coach, karate instructor, and nickel prospector. He is presently a visiting professor at the University of Florida.